Giuochi mathematici by Piero di Niccolò d'Antonio da Filicaia

Medieval and Renaissance Authors and Texts

VOLUME 30

The titles published in this series are listed at *brill.com/mrat*

Giuochi mathematici by Piero di Niccolò d'Antonio da Filicaia

An Edition with an Introduction

By

Luca Palmarini
Roman Sosnowski

BRILL

LEIDEN | BOSTON

The research upon which this publication is based has been supported by National Science Centre of Poland (grant agreement numer 2016/21/B/HS2/00744).
This publication was funded by the program Excellence Initiative—Research University at the Jagiellonian University in Krakow (Flagship Project 'European Heritage in the Jagiellonian Library: Digital Authoring of the Berlin Collections').

The Library of Congress Cataloging-in-Publication Data is available online at https://catalog.loc.gov
LC record available at https://lccn.loc.gov/2025016246

Typeface for the Latin, Greek, and Cyrillic scripts: "Brill". See and download: brill.com/brill-typeface.

ISSN 0925-7683
ISBN 978-90-04-72797-7 (hardback)
ISBN 978-90-04-72798-4 (e-book)
DOI 10.1163/9789004727984

This book is printed on acid-free paper and produced in a sustainable manner.

Printed by Printforce, the Netherlands

Contents

Figures and Tables

Figures

Tables

Introduction

CHAPTER 1

Piero di Niccolò d'Antonio da Filicaia and His Links with the Medici Family

1 Piero's Biographical Profile

Despite working during a crucial time for Florentine culture, the author is hardly remembered or known. The treatise was written between 1515 and 1516, around the time when Francesco Guicciardini, Niccolò Machiavelli, Ludovico Ariosto, and Pietro Bembo were active, in other words, in what can be considered the golden age of literary culture in Italy.

Piero di Niccolò d'Antonio da Filicaia was not a member of the elite group of intellectuals involved in Florentine political and cultural life. However, a relative on his mother's side was Bartolomeo Scala, a statesman, historian, and man of letters who had collaborated closely with the de' Medici family. His maternal ancestor served as Filicaia's motivation as well as a point of contact to win the goodwill of Giuliano de' Medici, duke of Nemours and Pope Leo's brother. Piero's father, on the other hand, came from a prominent line of Florentine copyists, including his great-grandfather Antonio da Filicaia (Bianco 2003, 238).

Altogether, in more favourable circumstances, this might have enabled him to enter the Medici *milieu*. Some sections of the *Giuochi mathematici* attest to his connections and his willingness to use them in order to gain access to the Medici family and to the political favours that this entailed. The premature death of Giuliano de' Medici in March 1516 was probably the direct reason for these expectations being frustrated.

Although Filicaia is absent from both the literary and political histories of Florence, we nevertheless have access to a variety of information about him because of the thoroughness of the Florentine archives. The family and the years of his stay in Borgo Sansepolcro are well outlined in Ulivi's reconstruction (Ulivi 2013). Piero, son of Niccolò and Maria Battista Scala, was born on 21 March 1489. In the 1490s, the young Piero was introduced to the study of Latin and vernacular writing.

As our Piero recalls, his maternal aunt Alessandra,[1] a sophisticated poetess who also wrote Greek verse and was courted by Poliziano, then wife of Michele

1 Alessandra Scala (1475–1506), was a Florentine poetess and scholar. She wrote her poetry in Greek and was widely known and admired for her scholarly advances.

 | DOI:10.1163/9789004727984_002

Marullo, oversaw at least a portion of his education (Ulivi 2013, 239) as long as he lived in Florence. The family moved to Borgo Sansepolcro after 1501. The young Piero da Filicaia resided in Borgo and finished his studies there under the guidance of Friar Pascasio, a monk who belonged to the same convent as Luca Pacioli, the famous mathematician who also played an important role in Piero's activity.

Piero remained at Borgo Sansepolcro until 1516, according to several documents (Ulivi 2013, 239). The author of *Giuochi mathematici* returned to Florence in 1516 and married Lena degli Albizi in the same year. Most likely a merchant, Piero held modest roles both inside and outside of Florence (Ulivi 2013, 241). His presence and activity in Florence in the 1520s are confirmed by a number of notarial documents discovered by Ulivi (2013). Regarding the date of Piero's death, the Florentine historian writes as follows:

> Apprendiamo dunque che Piero morì attorno ai quarant'anni tra il 21 novembre 1528 ed il 4 luglio 1531. La sua assenza nei "Registri dei morti" di Firenze fa pensare che il Da Filicaia possa essere scomparso durante la pestilenza del 1528, oppure che sia deceduto altrove, forse durante un incarico che lo aveva portato lontano dalla città natale.
>
> We thus learn that Piero died around the age of forty between 21 November 1528 and 4 July 1531. His absence in Florence's "Registers of the Dead" suggests that Da Filicaia may have passed away during the plague of 1528, or that he died elsewhere, perhaps during an assignment that had taken him far from his native city.

Perhaps an explanation on this last point can be found in Benedetto Varchi's *Storia fiorentina*. Varchi (1843, 2:297 and 451) again cited a certain Piero da Filicaia in 1529 and 1530, who was called to represent some citizens in the challenging circumstances of the fight with Charles v in 1529 and 1530.

Although there is no guarantee that this is our author, accepting the identification forces us to adopt a more precise date for his passing. This could have happened between July 1530, when he was last mentioned in *Storia Fiorentina*, and 4 July 1531. There is no record of any more literary or scientific activities other than his young efforts (the treatise on mathematical games and the *Trionfo* in honour of the Medici family); thus, it is obvious that his adulthood years distracted him from his studies.

Piero came to know and frequent Luca Pacioli, who often returned to the Franciscan monastery during his years in Borgo Sansepolcro, where Piero's grandfather and father held important positions during those years. He also studied and discussed the contents of the famous mathematician's treatise *De viribus quantitatis* with Pacioli. Several documents, many of which were rediscovered by Ulivi, confirm Pacioli's contacts with Piero Filicaia. In 1511, Piero is named as one of Pacioli's executors, and in 1512, a dispute over Filicaia's restitution of three chests left by Pacioli arises, as previously mentioned by Taylor (1942, 374–375) and now described in detail by Ulivi (2013, 243–253). Initially, Pacioli's property had been seized due to alleged non-payment of debts (17 florins) to Piero da Filicaia. The incident appears to have caused friction between them, with Pacioli denying owing any money and contesting the seizure of his property. To resolve this conflict, officials from Sansepolcro were involved to verify the facts and, if necessary, push for restitution of the property to Pacioli. This happened on 26 April 1513, when these officials also allowed the Captain of Sansepolcro to examine all witnesses. The dispute continued for several years. On 10 March 1516, a majority of the friars of the convent confirmed Filicaia's version of the situation. They stated that Pacioli had regained possession of his property, but only after it had been kept for some time with Maestro Santi, possibly during one of Pacioli's numerous absences from Sansepolcro. By the time this dispute was documented on 10 March 1516, Luca Pacioli was no longer present in the Franciscan convent and perhaps not even in Sansepolcro. He would die a little more than a year later, between April and October 1517.

The stay in Borgo Sansepolcro was thus crucial for the crystallization of Filicaia's mathematical interests, as well as for the writing of the treatise. Pacioli's influence on Filicaia's interest in the mathematical sciences is undeniable. Filicaia most likely refers to the texts in the Franciscan monastery library or the books entrusted to him by Pacioli: "ho visto, lecto e rivoltato molti auctori sopra tale materia parlanti" (f. 1^v). Piero gained a better understanding of the fact that "scientia mathematicha è infallibile" as a result of them.

His literary output is little known. He appears to have left no other works besides the treatise in question, two sonnets, and a poem about the festive procession in Florence after the election of Medici Pope Leo X.[2] As a result, the only sources for reconstructing the biographical story are a few mentions in *Giuochi mathematici* and documents from Florentine archives.

As this brief biography of the author indicates, literary activity was not central to his life. The only larger work for which we have any record is the treatise on mathematical riddles. The motivation for writing the treatise is clear from

2 This is the *Trionfo* which Piero Filicaia dedicated to Lorenzo de' Medici, son of Piero (preserved in ms. II.IV.310 of the Biblioteca Nazionale Centrale di Firenze, ff. 172–178, dated 3 April 1513).

the dedications and the setting. It is the work of an ambitious, well-educated young man with a special interest in mathematics who wanted to gain the favour of the lords of Florence by sending a valuable and welcome gift to Giuliano de' Medici. It is difficult to say whether the treatise had the desired effect; most likely not, because there are no further traces of our author's particular familiarity with Giuliano's *milieu*.

Piero was most likely deprived of such opportunities by Giuliano's untimely death (he died shortly after the treatise was written on 17 March 1516). In the grand scheme of Piero da Filicaia's life, his literary endeavour is pure *otium*, as opposed to his usual activity of *negotium*, or civil engagement. It is a work of youth, the result of youthful interests as well as external influences such as Paciolian ones. The subject also fits this viewpoint perfectly. It is *otium* of high quality, entertainment accessible to refined and educated people, and a mathematical leisure that combines playful and scientific elements.

2 The Dedicatee: Giuliano de' Medici

Piero da Filicaia dedicates his treatise to Giuliano de' Medici, Lorenzo's son and Leo X's brother. He was "come mecenate intelligente ed amico generoso"[3] whom "Leonardo, Raffaello, Michelangelo, Baldassare Peruzzi, Fra Giocondo apprezzarono"[4] (Sticco 1940, 143) and some of the Renaissance's most interesting minds dedicated their works to him (Fra Giocondo's second edition of Vitruvius's translation, Filippo Calandri's *Arithmetic*), or wanted to (Machiavelli).

He appears as a character in many of the most important early 16th-century texts, where he is "gentiluomo colto, raffinato, dal gusto perfetto, dalla conversazione sempre piacevole"[5] (Sticco 1940, 141). He was well-liked by those who knew him personally, both in his years of exile and later, as master of Florence, by rulers and friends, by commoners and refined men of letters.

2.1 *Youth*

Giuliano was born in Florence on 12 March 1479, the third son of Lorenzo di Piero de' Medici and Clarice Orsini. His older brothers, Piero and Giovanni, were destined to succeed their father and to pursue an important ecclesiasti-

3 Transl.: intelligent patron and generous friend.

4 Transl.: Leonardo, Raphael, Michelangelo, Baldassare Peruzzi, Fra Giocondo admired.

5 Transl.: a genial and sympathetic learned gentleman, refined, of perfect taste, of ever-pleasant conversation.

cal career, respectively. Their lives changed when Lorenzo il Magnifico died in 1492, but before that time they had a happy and interesting childhood. In the first years of his life, Giuliano remained under the tutelage of canon Matteo Franco, who in 1485 was a kind of elementary teacher in the Medici household and who in a letter to Piero da Bibbiena described Giuliano in the following words: "freschellino com'una rosa; gentile pulito e nettolino come uno specchio; lieto e tutto contemplativo con quei occhi …"[6] (Medici 1939, VIII).

After the initial period spent with his tutor Franco, together with Giovanni and his cousin Giulio he took lessons with Bartolomeo da Pratovecchio, Bernardo Michelozzi, and Gregorio Spoletino. We find several hints about his character as a young man in some Florentine chronicles. On the one hand, Bartolomeo Cerretani reports that his father Lorenzo said: "aveva uno filgliolo armigero, questo era Piero, uno buono, questo era il chardinale, uno savio, questo era Giuliano"[7] (Medici 1939, IX), and on the other, Piero Parenti, an opponent of the Medici party, recalls an episode in August 1493 in which Giuliano, only 15 years old, lost a finger in a dispute over a woman (Parenti 1994, 58). Parenti later describes Giuliano as "più che il dovuto ardito e voglioso"[8] adding that "una notte tagliato a pezzi non fussi, per il che scandolo a generare s'avessi"[9] explaining the October 1494 departure with his brother Giovanni for Rome (Parenti 1994, 110).

Because the eldest brother, Piero de' Medici, was unable to wield power in Florence, Giuliano and his brothers were exiled two years after their father's death. He lived then in exile until 1512, when the peninsula's political balance shifted again and Medici power was restored in Florence under Pope Julius II. His longest and, arguably, best years were spent at the court of Urbino (1503–1512).

The young Giuliano was a fascinating character, more endowed with the talents of his father, Lorenzo de' Medici, than his brothers. Magnanimous, loyal, and with literary interests, he embodied the ideal of the courtier in the eyes of his contemporaries. In battle, he demonstrated his valour and loyalty. He demonstrated this more than others in the episode of 1499, when he attempted to enter Florence with troops gathered with the assistance of the Venetians. After the initial successes of the army aligned with the Medici and the occu-

6 Transl.: fresh as a rose; gentle, clean, and neat as a mirror; cheerful and entirely contemplative with those eyes.

7 Transl.: He had a son-at-arms, this was Piero, a good one, this was the cardinal, a wise one, this was Giuliano.

8 Transl.: more than due bold and wanton.

9 Transl.: He was almost cut to pieces one night, which would have caused a scandal.

pation of Bibbiena, which was taken by deception, there were the hard times, when the soldiers, unpaid, began to desert the camp. Despite the desperate situation, even though the Duke of Urbino had withdrawn, declaring himself ill, Giuliano acted nobly and loyally by refusing the safe conduct offered by Vitelli, the Florentines' leader (Medici 1939, XVIII). He stayed with his own army until the fighting ended, despite the fact that he had no chance of winning. Giuliano then travelled to Venice to plead with the republic for assistance for the Medici cause. This had no political impact, but it did allow him to reconnect with Pietro Bembo, whom he had known since childhood (Medici 1939, XIX).

His life from 1494 onwards was marked by a series of struggles and growth in the shadow of exile. We find references to various episodes of these struggles in the chronicles describing Florence at the end of the 15th and the beginning of the 16th centuries. In 1495, through El conte Ugo della Gherardesca and the lord of Piombino, efforts were made to reintegrate Giuliano di Lorenzo de' Medici, the former's son-in-law, back into Florence (Parenti 1994, 209). By September 1498, Giuliano ventured to Florence, managing to take Marradi, but was eventually forced to retreat through the Lamone Valley back to Faenza (Parenti 1994, 299). Then, in December 1500, Giuliano, collaborating with Duke Valentino, again tried to restore Medici rule (Parenti 1994, 406). In November 1501, they once more sought to mount a comeback, this time supported by France. Yet, it seemed like a French threat against Florence rather than any true support for the Medici, since the French did not show any strong inclination towards the restoration of Medici rule. These initial years of exile were primarily characterized by diplomatic and military manoeuvres, always fuelled by the anticipation of a swift return home. High expectations were pinned on France, and more specifically on Louis XII's court, where Giuliano was a frequent guest. Florentine ambassadors including Francesco Soderini, Luca Antonio degli Albizi, and Biagio Buonaccorsi were present during one of these visits by Giuliano (Buonaccorsi 1999, XV).

From September 1501 to July 1502, the Florentine messengers were in France and "vi consumarono in vano circa a otto mesi senza avere mai una buona parola, anzi ributtati sempre con modi villani dal re, dal Roano e da tutta la corte, e fatto in presenzia loro e date lunghe audienze a Giuliano de' Medici"[10] (Guicciardini 1931, 217). This highlights the warm reception Giuliano received at

10 Transl.: spent about eight months there without ever receiving any encouraging response, instead always met with dismissive attitudes by the king, Roano, and the entire court, who, in their presence, granted Giuliano de' Medici lengthy audiences.

the French court, yet this affection did not translate into any tangible political outcomes. The French king was burdening Florence with significant economic demands and may have been using the threat of supporting Giuliano, whom the French court held in high regard, to deter Florence from resisting.

2.2 *At the Court of Urbino*

The *Dizionario Biografico degli Italiani* effectively summarizes the central period of Giuliano's life in the following words:

> La morte di Piero de' Medici, nel dicembre 1503, segnò la fine delle speranze di restaurazione medicea e fu un grave colpo personale per il M., il quale da quel momento visse prevalentemente a Urbino presso il duca Guidubaldo (morto nel 1508) e poi presso il suo successore, Francesco Maria Della Rovere, coltivando i piaceri letterari e mondani di una corte raffinata (Tabacchi 2009).
>
> The death of Piero de' Medici in December 1503 marked the end of hopes for Medici restoration and was a serious personal blow to M., who from that time on lived mainly in Urbino with Duke Guidubaldo (who died in 1508) and then with his successor, Francesco Maria Della Rovere, cultivating the literary and worldly pleasures of a refined court.

Giuliano spent his happy years in Urbino, sharing them with the most illustrious figures of the time who built the myth of Urbino, a promised land with an ideal court in an ideal city. He was completing his education there, and he was able to indulge in

> al godimento intellettuale che la corte di Guidobaldo sapeva dispensare così nobilmente a quanti avevano innato il culto del bello. Quivi raggiunse quella squisita raffinatezza di modi e di parola che lo fece apparire "uomo più presto da corte che da guerra", più adatto cioè ad essere il modello dell'elegante conversatore e del perfetto cortigiano che del guerriero. (Medici 1939, XXVII)
>
> the intellectual enjoyment that Guidobaldo's court knew how to dispense so nobly to those who had an innate cult of beauty. There he attained that exquisite refinement of manners and speech which made him appear "a man sooner of court than of war," that is, more fitted to be the model of the elegant conversationalist and perfect courtier than of the warrior.

His friendship with Count Baldassare Castiglione stems from these years. Because they were together at the ducal castle in Urbino during a critical period in their lives, there was never an exchange of correspondence between them; however, Baldassare mentions Giuliano in some letters (Castiglione 2016) to his mother, Aloisia Gonzaga Castiglione. Letters 108, 109, 112, 128, 134, 144, 147, and 153 are included. In particular, in 147 (10 Jan. 1509) and 153 (26 Feb. 1509), he mentions the failure of his marriage project with Clarice de' Medici, who went on to marry Filippo Strozzi at the behest of Giovanni de' Medici, despite Giuliano's opposition, who wanted to keep promises made to his friend Castiglione. Baldassare himself points out (letter 153):

> Di quello che la M.V. ha preso così a core, veramente non senza gran ragione, pur non si po' mo' tornare indrieto: forsi che è per il meglio. Da me non è venuto il difetto, e la M.V. lei stessa sa le parole che li disse il Mag.co. Io poi non cessava ogni dì de racordarlo; ma in effetto il Mag.co in questo caso non ha colpa, e so certo che a lui sommamente è dispiaciuto.

> Of what the M.V. took so to heart, truly not without great reason, though one cannot now go back *indrieto*: perhaps it is for the best. The defect did not come from me, and Your Excellency herself knows the words that the Magistrate said to him. I then did not cease every day to remind him of it; but in effect the Magistrate in this case is not to blame, and I know for certain that he is supremely sorry.

We can be certain of Giuliano's loyalty, which he demonstrated on numerous occasions and was valued by his friends, including Castiglione, in keeping with the ideal of the perfect courtier, which Giuliano seemed to embody more than others during those happy years in Urbino.

The other nucleus of Castiglione's letters in which Giuliano is mentioned revolves around Francesco Maria della Rovere, duke of Urbino, whose emissary Castiglione was in Rome. With Giuliano beset by illness and no longer able to oppose the aims of his brother Leo X and nephew Lorenzo, the plan to occupy the duchy of Urbino took hold with the excommunication of 1 March 1516 (letters 304, 305, 306). In particular, in letter 306, written on behalf of Francesco Maria to the College of Cardinals, the hospitality given by the dukes of Urbino to the pope's brother is recalled. Giuliano's name occasionally appears in Count Castiglione's letters later on, when referring to Giuliano's illegitimate son, Ippolito,[11] or in recollections of past times (1044, 1434, 1434, 1456).

---

11 Ippolito de' Medici, son of Giuliano and Pacifica Brandani, a noblewoman from Urbino, who later became a cardinal.

Without a doubt, the Urbino sojourn was Giuliano's period of maturation. It allowed him to lead a dignified life in the company of figures such as Bembo and Castiglione, who would shape many of the cultural aspects (literary, linguistic, and customs) not only of Italy but also of Europe. Giuliano, along with others then housed at the Urbino court, maintained the comparison and also became part of the literary myth of Urbino.

This also allowed him to thoroughly investigate his own soul and aspirations, both literary and political, and prepared him for future challenges when, along with his brother, Pope Leo X, and the trusted Bibbiena, he was to decide the fate of the Apennine peninsula (alas, for a very short period). He was so grateful to the lords of Urbino (especially Elisabetta Gonzaga) for their hospitality that he put political priorities on hold in 1515–1516 and opposed his brother's ambitions for Urbino for the de' Medici family (the occupation became effective after Giuliano's death).

2.3 *Giuliano as a Literary Character and Poet*

Giuliano's fortune as a literary character is inextricably linked to the Urbino period. He is described in Castiglione's *Tirsi* as follows:

> Venne d'Etruria un altro in questi monti,
> Saggio o dotto pastore in ciascun arte.
> Non son piagge qui attorno o rivi o fonti
> Che non intendan le sue lode sparte;
> Ma temo assai che prima il sol tramonti,
> Ch'io possa dir di lui pur una parte;
> Questo cantò con amorosa voce:
> "Se fosse il passo mio così veloce".
>
> Came from Etruria to these mountains / [another one,] A shepherd wise and learned in every art / There are no shores, rivers, or springs around here / That do not understand his scattered praises; / But I am afraid that before the sun sets, / I will only be able to say a part of him; / This he sang with a loving voice: "If my pace was so swift."

According to Fatini, although the friendship with Castiglione had not reached the intimacy of that with Bibbiena (a childhood and study friend) and Bembo, it was firm because it was also based on an appreciation of the courtier-cavalieresque ideal that Castiglione would praise in the *Cortegiano* and that Giuliano felt was his own aspiration (Medici 1939, XXIX–XXX).

Much more significant is his presence and role in his friend Pietro Bembo's major work. Giuliano defends the vernacular in the *Prose* (published in 1525), as "a language closer to us, more natural and proper". He and others try to persuade Ercole Strozzi, a humanist-Latinist, of the Italian vernacular's maturity. However, in opposition to Carlo Bembo who praised the 14th century, Giuliano represents the party of the 16th-century Florentine. He argues that

> le scritture, sì come anco le veste e le arme, accostare si debbono e adagiare con l'uso de' tempi, ne' quali si scrive, con ciò sia cosa che esse dagli uomini, che vivono, hanno ad esser lette e intese, e non da quelli che son già passata. (*Prose*, I, XVII)

> writings, as well as clothes and arms, must be approached and adapted to the use of the times, in which they are written, so that they may be read and understood by men who live, and not by those who have already passed away.

He also elaborates on the effective comparison, which is then countered by Carlo's reasoning, a spokesman for Pietro himself, about the absurdity of using a language that is no longer spoken:

> Perché, sì come voi e io saremmo da riprendere, se noi a' nostri figliuoli facessimo il tedesco linguaggio imprendere, più tosto che il nostro, così medesimamente si potrebbe per aventura dire, che biasimo meritasse colui, il quale vuole innanzi con la lingua degli altri secoli scrivere, che con quella del suo. (*Prose*, I, XVII)

> For, just as you and I would be blamed if we taught our children the German language rather than our own, anyone who prefers to write in the languages of other centuries rather than his own could be blamed for the same reason.

In the third book, he is entrusted with the difficult task of expounding vernacular grammar in a precise, elegant manner while avoiding technicalities and linguistic pedantry. Leaving aside the question of how much of the real and how much of the literary were involved in the creation of Giuliano's character in the *Prose*, it appears clear that Giuliano is entitled to a prominent place and that "la parte che il Magnifico ha in questo libro è stata dettata all'autore non solo dalla profonda amicizia per lui, ma anche dal ricordo delle discussioni e dal riconoscimento della competenza che in essa portava"[12] (Medici 1939, XXVII).

12 Transl.: the part that the Magnifico has in this book was dictated to the author not only

Finally, in the *Cortegiano* dialogues, he plays an important, although perhaps not central, role. He became a "pittore esperto della perfetta donna di corte"[13] (Medici 1939, XXVIII) and defended women by drawing on their idealized profile against the misogynistic positions of Gaspare Pallavicino and Ottaviano Fregoso.

Celebrated as a character by many of his contemporaries and appreciated as a poet in his own time, he was less fortunate with posterity, always remaining in the shadow of his father Lorenzo, who was certainly better known, livelier, and more original. Both the editor of his literary output, Giuseppe Fatini (Medici 1939, CXXIX–CXXX), and the literary historian Maria Sticco (1940, 145) agree that Giuliano's own voice, that is, his poems, lack elegance and intensity.

Somehow, there is more originality in the poems with the accompanying theme of jealousy, and in those in which a genuine and deep love for the homeland Florence shines through (Sticco 1940, 147–151). This critical judgement may be mitigated or modified when the new edition of Giuliano's rhymes announced by Giacomo Vagni is published (Vagni 2016).

2.4 *The Triumphant Return to Florence*

After a long exile, the Medici returned to Florence following the Congress of Mantua (1512). On 13 September, Pietro Bembo congratulated Giovanni de' Medici on this historic occasion, writing:

> Magna me laetitia, magna voluptate affectum esse de tuo, deque Iuliani fratris tui tam secundo in patriam reditu, de hac in vos tam propensa Deorum hominumque voluntate, etiam tacente me, puto tibi esse perspectissimum. / Idibus Septemtribus. MDXII. Roma. (Bembo 1987, 62)
>
> I think it is clear, even if I am silent, that I feel great joy and great pleasure on your and your brother Julian's return to the fatherland for the second time and that the will of gods and men is so favourable to you. On the Ides of September, 1512, Rome.

Giuliano took over the city government without retaliation. There were no consequences, as some in Florence had feared. On the contrary, Giuliano, who was universally known "per la mitezza d'animo e per il desiderio di vivere tranquil-

---

by deep friendship for him but also by the memory of the discussions and the recognition of the competence he brought to it.

13 Transl.: expert painter of the perfect court woman.

lamente"[14] (Medici 1939, XXXV), attempted to adapt to the lifestyle and even dress of his fellow citizens. Giuliano founded the *Compagnia del Diamante* in November with a group of young nobles to commemorate the return of the good times and to revive the memory of the famous festivities organized by Lorenzo de' Medici 20 or 30 years earlier. The *Compagnia del Diamante* organized carnival games and parties, complete with floats, masquerades, and triumphs. Thus began the carnival of 1513, in which people appeared to have returned to the carefree and merry years of Lorenzo Il Magnifico. Giuliano's nephew, Lorenzo di Piero de' Medici, organized the *Compagnia del Broncone*. The two companies, *del Diamante* and *del Broncone*, engaged in a battle of carts (painted by Jacopo Pontormo) and to the tune of specially composed songs (Medici 1939, XL–XLI). The year 1513 thus began with triumph and ended with even brighter prospects when, following the death of Pope Julius II, Giovanni de' Medici was elected pope under the name Leo X. It would be a mistake to focus solely on luck and chance when thinking of these events. Certainly, favourable circumstances were required for this significant change in the brothers' lives, but it also required great skill on the part of Giovanni, his trusted secretary Bernardo Dovizi da Bibbiena, and his cousin Giulio de' Medici. Giuliano's brother's election to the papal throne provided him with new opportunities as well. Years of turmoil followed, filled with projects and plans, some of which were realized while others were not. What cannot be disputed is Giuliano's importance in Leo X's political plans, as he built the de' Medici family's future fortunes through the papacy, as was customary at the time. Some of Bibbiena's letters to Giuliano attest to this. On 16 February 1515, Bibbiena reminds Giuliano of his importance to the de' Medici household:

> Ricordatevi che dopo N[ostro] S[ignore] ognuno mira a voi, essendo la persona vostra quella, nella quale i pensieri, i concetti, e i disegni del Papa si specchiano. Et anche vi ricordo, che ogni opera, e attion vostra non è considerata e notata manco, che quella di N.S. (Ruscelli 1562)
>
> Remember that everyone looks at you after Our Lord because you are the one who reflects the pope's thoughts, concepts, and designs. I also remind you that your work and actions are not less important or noticed than Our Lord's.

14 Transl.: for his mildness of mind and desire to live quietly.

And again on 11 March 1515:

> Sua Santità è piu che mai ardente al vostro bene, et honore, e sono certo lo dimostrerà con effetto. (Ruscelli 1562)
>
> His Holiness is more committed than ever to your good and honor, and I am confident he will demonstrate it with effect.

His marriage to Filiberta of Savoy, who was related to French King Francis I, took place in 1515. On this occasion, the king bestowed the title of Duke of Nemours on Giuliano, and the pope appointed him Captain General of the Church (Tabacchi 2009). "Il Magnifico si scelse come motto allusivo alla sua fortuna sei lettere disposte a mo' di triangolo; GLOVIS, che lette a rovescio significano Si volg(e)"[15] (Medici 1939, LXXI). Because the motto GLOVIS appears on the title page and Filiberta's marriage to Giuliano is described in the introduction to the fourth part of the work, this information is extremely helpful in dating Filicaia's work.

However, the year 1515 also marked the beginning of Giuliano's serious illness, which would lead to his death on 17 March 1516. His departure threw a wrench in the family's dynastic plans. Soon after, the pope's nephew, Lorenzo di Piero de' Medici, died, effectively ending the family's secular ambitions. Nonetheless, Giuliano's funeral served to highlight the family's strong ties to Florence, with an explicit call for reconciliation under Medici leadership. The oration of Chancellor Marcello Virgilio Adriani clearly demonstrates this intent (McManamon 1991, 1). Giuliano's non-dominant personality, keen attention to civic forms, and princely magnanimity made him as popular among Florentines as he had been among princes, sovereigns, and courtiers wherever he went.

Giuliano's curiosity and magnificence toward artists and scholars is an intriguing aspect of his character and actions. Giuliano was interested in new geographical discoveries and the new eastern regions that were opening up to travellers in the early 16th century, as evidenced by Andrea Corsali's 1515 letter-report sent from Cochin, India:

> Illustrissimo Signor, non potendo manchar a V.S. di quanto le promissi nel partirmi di costì, ho voluto farle questo poco di discorso p(er) darle notitia del successo del mio viaggio d'India. (Ramusio 1550, 192^{v})

15 Transl.: The Magnificent chose for himself as a motto alluding to his fortune six letters arranged in the shape of a triangle; GLOVIS, which means Si volg(e).

> I wanted to make you this little report, Illustrious Lord, in order to give you notice of the success of my journey from India, because I didn't want to fail what I promised you in departing from there.

3 The Reasons for Filicaia's Dedication to Giuliano de' Medici

The dedication is particularly interesting because it provides us with important information about both Filicaia and Giuliano de' Medici, as well as his relationship with mathematical science. The dedication is the written expression of an individual's but also a collective's desire to define the contours of the relationship that the gift of a work (in this case, a manuscript) establishes between its author, who is determined to donate his work, and the dedicatee, who happens to receive it. In cases like the *Giuochi mathematici* treatise, the dedication serves as important evidence of the relationship between intellectuals and power, as well as the potential consequences of such a gesture. It is both an explicit and codified tool of inquiry, to be interpreted in accordance with the variations that individual dedications set out to create a pattern in normative assumptions (Villa 2010). In precisely codified language, the dedication functions as a proposed contract with no legal value but a profound moral value of loyalty and servitude on the part of the author, who wishes to join Julian's court.

Filicaia, like Niccolò Machiavelli and others, followed a literary convention in the 15th and 16th centuries. Authors offered their works to political leaders in exchange for protection and "grace" (Rinaldi 2001). The writer is thus the first to act, to take the initiative, and through the gift, he communicates his encomiastic intentions. We have a truly extensive repertoire of Renaissance dedications; they contain many variations of the relationship between powerful man and intellectual, developed through an elaborate strategy with the primary goal of valorizing and thus disseminating works. Such dedications are logically tailored to each recipient and a specific political situation. Rinaldi (2001) writes openly about a literature governed by the regime of offering and dedication, in which the historiographical genre dominates, which is best suited to legitimizing the power of a given ruling house; however, it is clear that such encomiastic elements can also be found in other genres, as Filicaia's treatise of demonstrates.

Undoubtedly, until the 16th century, when Luca Pacioli and Niccolò Tartaglia realized the need to create a link between these two domains, the academic and practical domains hardly ever interacted (Pietrini 2020). These phenomena surely affected the content of many dedications.

It is obvious that, from the standpoint of the current work, what most interests us is related to Giuliano's cultural interests and his relationship with mathematics. The treatise by Piero di Niccolò da Filicaia is not the only evidence of Giuliano's interest in the subject. Already in his treatise *Aritmetica*, Filippo Calandri mentions the young Giuliano's passion for mathematics, dedicating his work to him (Calandri 1491):

> Philippi Calandri ad nobilem et studiosum Julianum Medicem de arithmetica opusculum.
> Considerato, nobile et studioso Juliano Medice, quanto sia utile, anzi necessaria, la scientia arimethrica al comertio humano et maxime a quegli che exercitano la mercatura, di che la ciptà Fiorentina sanza controversia fra l'altre tiene il principato, et veduto la grata et celebre audientia degli studiosi adolescenti fiorentini in questa mia giovenile età, m'è paruto conveniente le cose da me udite a llor maggiore utilità sotto brieve compendio ridurre et quelle secondo lo stile fiorentino non con piccola mia fatica, per le multiplice difficultà che agli pressori ocorrevano, per più comodità fare imprimere. Il che avendo per la divina gratia absoluto et volendo di già questa mia operetta andare in luce, accioché con maggiore gratia et auctorità vada, a te, Juliano Medice, la dirizo et dedico, che se' di tale scientia fra l'altre studioso, et secondo l'optimo costume de tua antecessori della publica utilità et honore amatore et defensore, la quale, se da te, come spero, sarà aprovata, mi sia stimolo di maggior cose a tentare et più artificiose. Vale.

> A small work on arithmetic to the noble and studious Giuliano de' Medici by Filippo Calandri.
> Consider, noble and scholar Giuliano, how useful, indeed necessary, is arithmetic to human industry and especially for those dealing in merchantry, in which subject the city of Florence is without doubt a champion, and seeing the gracious and famous audience in the young Florentines in my young age it seemed to me convenient to reduce the things heard by me into a brief compendium and this according to the Florentine style, for their greater usefulness. Because of the numerous difficulties that the pressors required, I expended considerable effort to impress them. Having been absolved by divine grace, and already desiring that this little work of mine be brought to light, so that it may come to you, Giuliano, with greater gratitude and authority, I direct and dedicate it to you, Giuliano, who are a scholar of such knowledge among other scholars, and according to the excellent custom of your ancestors

> of public utility and honor, lover and defender, which, if approved by you, as I hope it will be an incentive to attempt more complex things. Vale.

When we consider the dedication, which attests to Giuliano's early interest in mathematics and the popularity in Florence of mathematical riddles, which constituted a learned but widespread form of entertainment, we are not surprised that our author seeks de' Medici's assistance. Piero's pro-Medicean stance, which was strengthened by the aforementioned family ties to this household, should also be considered, as well as Giuliano's well-known status as an art patron. During the years that the treatise was being written, Giuliano employed Raphael and Leonardo, as well as intellectuals and other figures who might have appreciated the work on mathematics.

From the very first words of Filicaia, we encounter the doctrine of the addressee, expressed in the initial encomium to Julian, rendered in traditional Latin forms of politeness, which, however, deliberately stand in stark contrast to the *fiorentino argenteo* of the rest of the text. The purpose—Filicaia by no means presents the work as unique; Calandri himself, for example, resorts to Latin for the formulas of welcome and courtesy and then switches to Florentine—is to give a universal character to the praise expressed towards the duke and, consequently, to the author himself: "Ad Magnificum ac Clarissimum virum Julianum Medicen Patritium generosissimum Florentinum ac Dominum suum colendissimum Petri Nicolai Antonii de Filicharia prefatio incipit foeliciter." Those who extolled the dedicatee inspire admiration and a desire for protection, which inevitably leads them to affix their signature and declare themselves the author of these words.

Filicaia's dedication to mathematical science goes far beyond that of Calandri, who limits himself to a practical and concrete assessment, calling this discipline "useful" and "necessary", recalling its use in "comertio humano", and in "mercatura". Mathematics, on the other hand, is to be exalted and honored above all other sciences because it possesses "vera et unica perfectione", which means it is infallible and comes closest to the celestial element, not only in its perfection but also in its progression to infinity, just as the most holy Trinity is infinite. And who better to deal with such infinite science so close to the divine than Giuliano de' Medici, who reigns precisely through divine intercession while also reigning by virtue of reason and intellect? Furthermore, the Duke of Nemours may appreciate the work dedicated to him as a profound connoisseur of aspects of philosophy; in fact, this is another theme that Piero emphasizes in his dedication, using quotations from thinkers such as Euclid of Megara and Aristotle.

The dedication suggests a shift away from the practical application of mathematics toward the delight, conviviality, and playfulness to which the perfection of that science can lead the intellect; an environment of which Filicaia aspires to be a part.

The introduction to the fourth part confirms Filicaia's desire to gain Giuliano's favour and become a member of his inner circle. It concerns a dreamlike vision of the wedding of Giuliano and Filiberta. The duke is shown seated on a golden throne resting on a cloud, while his consort is shown seated on a throne supported by two lions. This is also indirect evidence of the transformation of power then occurring in Florence. Filicaia's declaration that he wholeheartedly wishes to stay with them brings the happy atmosphere to a close. The preservation of republican appearances, so important even two generations ago, is not maintained in the immediate post-republican years. The Medici family's power is evolving toward a type of seigniorial power. Although many characteristics of the Medici family and Italian courts were common during the years to which we refer, it is still difficult to speak of an actual court (the word "family" is usually used).

In the edition of Matteo Franco's letters, there is a summary of the contents of an interesting 1492 document entitled "Ritratto della ispexa della gita a Roma" (marked MAP, F. CIV no. 57, cc. 580–585 of the Archivio di Stato di Firenze). It tells us that the Medici family, or the court, numbered around 50 people. Giuliano's family (or court) was equally numerous and similar in composition during his return to power in Florence. However, it was not a formal court, but rather the family and a few intimates among their fellow citizens.

Those who were part of it reaped many benefits, so it was many Florentines' ambition to serve the Medici, which also entailed, in practice, a communal, courtly life. It should be noted, however, that the very nature of Medici power was "republican", with Lorenzo or later Giuliano being formally "primus inter pares". The daily routine of life was provided not only by the exercise of power but also by amusements, festivals, and cultural interests.

There could be no shortage of mathematical riddles among the amusements of a circle of friends eager to satisfy the interests of their first citizen, the result of both mercantile culture and Florentine Neoplatonism with its worship of numbers. For the reasons stated above, Giuliano is the ideal dedicatee of a treatise that serves as a kind of manual of learned amusement imbued with a strong idealistic component. Furthermore, Giuliano de' Medici's interests encompassed not only mathematics but also related sciences. When Fra Giocondo, another family member, prepared the second, revised edition of Vitruvius's treatise *De architectura*, he dedicated it to Giuliano and wrote:

Virtutes tam liberales, quanto quam mechanicae tanto magis proficiunt perficiunturque, quanto apud eos tractantur, qui illarum delectationibus detinentur et indefesso sollertive studio eas persequuntur, ut de te audio, Juliane Medices vir optime et earum studiosissime, cuius fama cicrumfertur, quod nihil a patre et proavis tuis excellenti ingenio et maximis virtutibus praeditis degeneras. (Vitruvius 1513)

Liberal virtues, more than mechanical virtues, advance and reach perfection when pursued by those who are seized by their pleasures and follow them with tireless and strong passion. Such are the opinions I hear of you, Julian, man of the best and most diligent of them, whose fame spreads, because you are in no way inferior to your father and ancestors endowed with outstanding talent and great virtues.

CHAPTER 2

The Two Manuscript Witnesses

1 The Description of the Two Witnesses

So far, we have two witnesses to the text. The first is in the Berlin Collection at the Jagiellonian Library in Kraków, catalogued as Ms. Ital. Quart 48. It will also be referred to as **K** in the following lines. The second witness can be found in the Biblioteca Nazionale di Firenze with the signature Magl. CL XI 15; it will be abbreviated as **F.**

The witness Ital. Quart 48 will serve as the foundation for the edition, so it is presented first: K= Ital. Quart. 48 is a paper copy about 225 mm high and 145 mm wide from the Jagiellonian Library in Kraków (Berlin Collection), with 181 folios plus two flyleaves. Watermarks: P in shield surmounted by star, 58 mm × 38 mm (star 30 mm); Eagle in circle (= Piccard online, 042665: Como 1512). The manuscript was foliated in pencil during the Renaissance period; the folios containing the index are numbered in Roman numerals (II–VI); however, Arabic numerals are used for the foliation of the main text. The written surface measures 150 × 85 mm, with the text written on 21 lines.

The manuscript is embellished on f. 1^r with a frame with gilded floral motifs and the Medici arms at the bottom, surrounded by a diamond-set ring (Medici symbols); the inscription VIVA SENPER appears on the ribbon entwined with the ring. Also raised in gold on f. 1^r is the fiorite initial C; VIVA is written in gold within the field of the initial C. Other folios have simple initials. Illustrations of mathematical problems can be found on folios 166^r, 167^r, 169^r, 170^v, 172^r, 172^v, 174^r, while tables, drawings, and calculations can be found on folios 57^r, 82^r, 106^r, 109^r, 113^r, 113^v, 114^r, 114^v, 115^r, 115^v, 116^v, 126^v, 126^v, 127^r, 135^v, and 145^r.

The binding, from the old Medici library, is original and valuable. It is 227 mm high and 147 mm wide. The plates are made of wood and covered in green silk. The silk is heavily damaged. There are embroidered Medici arms in the centre of the plate, as well as studs at the corners and traces of the clasp, as well as the remains of the leather spine. The manuscript is now stored in a protective, rigid cardboard slipcase covered with marbled paper, which was made in the 19th century. On the plate is an oval label with the catalogue number (635 bis) and on the spine the Königliche Bibliothek label with the signature: Ms. ital. quart. 48. The auction label with the printed text explaining the contents and importance of the manuscript, correctly indicating the addressee and the date of completion (1510–1515), appears on the counterplate. The Berlin Library

 | DOI:10.1163/9789004727984_003

accession number and shelfmark are located in the centre of the counterplate. The auction label and information from the Berlin Staatsbibliothek's entry register indicate that the codex was purchased at the Franchi library auction in Florence (February 1885, catalogue 44, number 635 bis). Searches in old Medici catalogues of codices have yielded no results, making it difficult to determine how and when this manuscript was removed from the Medici collection (in all likelihood it must have been found there given the binding, unquestionably Medici). The first description of the manuscript was provided by Sosnowski (2012).

Let us now examine the Florence witness. **F** = Magliabechiano XI, 15; this is a paper copy about 207 mm high and 145 mm wide, 183 leaves in total (eight initial folios numbered with Roman numerals I–VIII with the index from I^r to I^v). Watermarks: P in shield surmounted by star, 58 mm × 38 mm (star 30 mm), up to f. 48; Eagle in circle (= Piccard online, 042665: Como 1512), from f. 49. As a result, these are the same watermarks found in the other manuscript. The work is foliated in ink up to f. 24, which has been continued with modern foliation in pencil. We have a modern foliation on ff. I–VIII. The ruling is by impression, and the written surface is 166 × 83 mm in size. There is evidence of several hands (cursive and chancery). Pen drawings can be found on 165^r, 166^r, 168^r, 169^v, and 171^r. Paper 35^v is blank, with erasures.

The binding dates from 1917 ("Ciabani Gino Legatore di libri"), the spine is vellum, and the inscription *P. da* FILICAIA *Giuochi matematici* appears on it. Annotation on the back counterplate: "Carte 175 con esatta numerazione; più 8 carte preliminari modernamente numerate in cifre romane delle quali sono bianche le carte V–VIII. Settembre 1917."

On the counterplate there is a tag with the shelfmark and provenance: *Marmi*. In the handwritten catalogue of the Biblioteca of Anton Francesco Marmi, in volume III of the catalogue devoted to manuscript works, compiled by Antonio Francesco Gori and Antonio Cocchi (Florence, National Library, *Fondo principale*, II.I. 276), we find, on paper 63, at number 62, the following description: "Libro detto Giuochi matematici, composto per Piero di Niccolao di Ant[onio] da Filicaia, dedicato al Magnifico Giuliano de' Medici, il di cui nome è stato cancellato, c. sec. XV, 4°, carta." A tag glued to the first folio of the manuscript reads: "Di mano di Piero di Niccolao d'Antonio da Filicaia libro dedicato al Mag co Giuliano de Medici figlio di Piero di Cosimo Padre della Patria che fu ucciso nel Duomo di Firenze ne' 26 Aprile 1478 nella congiura de' Pazzi." As Arrighi has already pointed out, the identification of Giuliano with Lorenzo Il Magnifico's brother is obviously incorrect. Giuliano de' Medici is the son of Lorenzo, duke of Nemours, born in 1479, as evidenced by chronology but also simply inferable from the text.

The author's hand could be the one that finishes the book (from paper 160). It makes frequent and authoritative corrections, leaves space for drawings, and is the same hand that writes the final two sonnets ("Petrus Filicharius Auctor librum suum alloquitur") (transposed later in K). However, in the absence of a comparison, such as a specific autograph by Piero da Filicaia, this is a speculative attribution. The manuscript is also described by Warren Van Egmond (1980) and catalogued in *Iter Italicum* (Kristeller 1963, 1:118). The description and edition of some excerpts can be found in Arrighi (1971). (Edits: ff. 1^{r-v}, 2^{r}–5^{r}, 159^{v}–163^{v}, 173^{v}–174^{r}, 174^{v}–175^{r}).

2 The Relationship between the Two Manuscript Witnesses

At this point, a brief comparison of the two *Giuochi* witnesses is required. While K is lavishly decorated, F has spaces for unfinished decorations. As a result, it is configured as "a bad copy" of K. This fact is also supported by the presence in K of an accurate index of the topics covered and by cross-references to individual pages at the start, which F does not include. Otherwise, the manuscripts are nearly identical. They have the same *mise en texte* and *mise en page*: both have the dedication to Giuliano de' Medici, duke of Nemours, and both have the same text arrangement (*Dedica*, *Epigrammi* di Fra Pascasio, text of the *Giuochi mathematici*, two sonnets by Piero di Nicolao da Filicaia).

Several corrections present in F are transposed in K, others are not. A sample of these is offered here:

- 1^{r} The title of the epigram is slightly different in F and K ("eiusdem Petri auctoris" versus "Petri auctoris infracripti");
- 3^{r} in F, "per virtù" was inserted in the margin, correction transposed in K;
- 4^{r} in F, "Prologus" was inserted in the margin, correction transposed in K;
- 9^{r} in F, "che avanza" in the margin is proposed for insertion, not transposed in K;
- 36^{r} in F, there is the correction "le figure" which corrects "le fractura", in K we find "le figure" thus transposed;
- 59^{r} in F "la regola" has been corrected to "lunghe" while "et questa somma dirai l'acozino insiemi" is deleted (repeating the previous line). In K there is correctly "lo tengha" instead of "lunghe" and no deletion is present because the text is already arranged;
- 78^{r} in F, a long correction can be observed. In K the corresponding text is correct, but there are some minor variants ("parteggerai" instead of "dirai");

- 117^{v} in F, we have the insertion of "razorraj per traverso". In K there is already "racorraj per traverso".
- 118^{r} in F, "essendo" has been corrected to "el secondo", already present in K. We have the insertion of "et tu dirai", already present in K;
- 160^{r} in F, "substenevano" is corrected to "substenuta", which is already "substenuta" in K;
- 164^{v} in F, the insertion of "fatto che" occurs, a correction also present in K (but with a different technique, cleaned up and overwritten with "fco"—"facto");
- 174^{v} in F, "sarai" is deleted, in K the lesson is correct, without "sarai".
- 175^{r} in F, we witness the insertion of "o", in K "or" is already present.

Considerations regarding paper are crucial for both witnesses. When creating the book, the copyists (or Filicaia himself) used the same paper for Magl. XI, 15 (= F) and Ital. Quart. 48 (= K). They begin with F (paper with watermark P) and gradually (or in one long sitting) copied it to K (the same paper as in c. 49 in F), or copied it to K after completing F. There are several corrections, cut papers, and alternating hands in F, implying that this is the first version that was later poured into what was to be the final one, i.e., K (Ital. Quart. 48). This hypothesis is further supported by the decorative layout, which is missing in F but present in K. Given that F and K use the same batch of paper and that, in any case, the timespan is limited, the layout of F and K must have been either contemporary or at least close to it (between the date when Giuliano receives the title of Duke of Nemours and the date when he dies). Another possibility is that F and K were copied independently from a common antigraph, but K was immediately chosen as a publication specimen while F was used for necessary revisions. The majority of F corrections are transposed in K, and the resulting lessons are correct in K.

As previously stated, these are two witnesses that were written at the same time on the same type of paper, both, we can assume, under the author's close supervision (and perhaps partially by his hand—unfortunately, parameters for comparison are lacking; however, we can reasonably assume that the corrections in the margin of F are by the author). Specimen K is a dedication specimen, written by one hand (unlike F), and thus the one corresponding to the author's will. As a result, we generally prefer the forms of K in the edition if they differ (in rare cases) from those of F. We use F to correct errors in K on occasion.

For all of the reasons stated above, the two manuscripts (F and K) can be defined as idiographs with a special relationship between them.

3 A Brief Glimpse into the History of Mathematical Riddles

Before proceeding with the analysis of the manuscript's content, a brief but necessarily superficial history of mathematical riddles is useful in order to place the edited text in a suitable context. Our task is philological rather than historical in nature. The publication of Filicaia's treatise, which represents one of the most significant milestones in the evolution of how mathematical riddle were presented, should aid in the reinterpretation of this history. However, it cannot replace the work of a historian.

As eminent historians have described the development of these mathematical tools with familiarity and detail, we do not claim to present a fuller version, but refer to the most important works in the field, here outlining only the background to Filicaia's Renaissance treatise.

We must begin by stating that reconstructing the history of recreational mathematics, which is practised and applied for enjoyment, is a challenging task. This is primarily due to the challenge of defining the boundary between serious, pure science, or strictly professional applications, and enjoyable, leisurely activities. We must likely acknowledge that mathematical works, which are not strictly theoretical but grounded in practical examples, frequently blend concrete, everyday problems with puzzles and invented questions, all presented in a manner that captures attention, stimulates curiosity, and provides entertainment.

A good attempt at presenting a scientific outline of the history of mathematical riddles was provided in Raffaella Franci's monograph on Alcuin of York (732–804), who is thought to be the author of the oldest collection of mathematical problems classifiable as riddles (written in Latin). Born into a wealthy family, most likely in York, he distinguished himself through study even as a young man. He soon enrolled in the cathedral school at York, founded by Archbishop Egbert, who, along with Albert, the school's master, recognized the rare abilities of their young pupil and assisted him in developing his talent. After about fifteen years, Alcuin was appointed rector of the York school. Under his leadership, the school gained a European reputation and saw a significant increase in student enrolment.

During one of his many journeys, Alcuin met Charlemagne in Italy in 781, who invited him to the Frankish court, which he accepted in 782. For several years, he led the Schola Palatina and was in charge of reorganizing education in Francia and throughout the empire on behalf of the emperor. He wrote on a variety of subjects, including grammar (also rhetoric and writing), theology, philosophy, history, poetry, and more. One of the consequences of Alcuin's reform was a growing awareness of the distinction between Latin and the languages actually spoken in the Romance area.

Alcuin's interest in mathematics was strictly didactic. He wanted to encourage his students to practice reasoning and calculations as a good teacher. The title of his book, "Propositiones ad acuendos juvenes", attests to this. Because of the content and the manner in which the mathematical problems were presented (as questions or riddles), this collection falls into the category of so-called recreational mathematics.

Such a co-presence of didactic motivations and light forms of question-guessing can be found in *abacus* treatises throughout the Middle Ages.

Since the *Liber abaci*, regarded as "one of the most influential books ever written" (Devlin 2013, 124) by Leonardo Pisano (Fibonacci), the presence of mathematical games has been relevant in any abacus. In the treatise, Fibonacci introduced Arabic numerals, which he referred to as "Indian" as well as the sign of zero, he also presented the divisibility criteria, square and cube roots, and other important topics. However, in addition to all of the theoretical issues raised, he devoted a portion of his work to mathematical puzzles with solutions. Among several problems (mostly involving the distribution of food and money), Fibonacci is well known for one. It's the rabbit population dynamics conundrum. Actually, he did not invent the problem (it is present in a different form among Indian mathematicians some centuries before Fibonacci), but the sequence of numbers that explains the solution to that question has become known as the "Fibonacci sequence".

As commerce and banking boomed, abacus books proliferated. A good merchant had to be able to keep accounts, but he also had to be able to resolve real-world problems in his business practice and deal with customers. This was why he required exercise books. The teaching of mathematics encompassed a wide range of problems in mercantile arithmetic and practical geometry, with the aim of engaging and intellectually stimulating pupils during the teaching process through mathematical puzzles and games (Ulivi 2008, 413). As Franci (1981, 21) notes, the goal was also to divert students' attention from tedious business applications. At the beginning of Chapter VI devoted to ("certi chasi dilettevoli e numeri"), Filippo Calandri wrote: "Ogni sano intelletto arebbe in fastidio non ragonando d'altri casi che di merchatantia, onde nello presente capitolo intendo mostrare alchuno chaso di dilecto e quali sono posti sopra numeri"[1] and, at the start of Chapter VIII (again devoted to riddles): "Io mi credo per lo chapitolo passato l'appetito è infastidito di tanto avere scripto di merchatantia, onde in questo presente capitolo intendo scrivere alcuno chaso di

1 Transl.: Every healthy mind would be annoyed not thinking about cases other than commerce, so in this chapter I intend to show some delightful cases which are based on numbers.

dilecto acciò che l'animo sia alquanto plachato".[2] Franci (1981, 21) recalls, citing Francesco Ghaligai's 1552 *Pratica di mercatura*, that the motivation for inserting the riddles in arithmetic manuals could have been playful, a pastime for long winter evenings.

De viribus quantitatis, written by Luca Pacioli, was the first modern treatise devoted entirely to ludic mathematics (though it was never published in print).

And here, chronologically, we place Piero di Niccolò d'Antonio da Filicaia's treatise. Right after Pacioli, written in the vernacular, like the Paciolian treatise, and never published in print.

After Piero da Filicaia, there is a long list of famous mathematicians who, in addition to studying, enjoyed amusing and entertaining themselves with puzzles and riddles. To stay in the Italian realm, it is worth mentioning Niccolò Fontana, also known as Niccolò Tartaglia (1499–1557), and Girolamo Cardano (1501–1576), two famous mathematicians who contributed significantly to 16th-century mathematical knowledge. However, mathematical games are only a small part of the content in their works.

When discussing the history of mathematical games, another significant figure should be mentioned: Claude-Gaspar Bachet de Méziriac (1581–1638). This French mathematician was the author of "Problèmes plaisants et délectables qui se font par les nombres", which was published in 1612. As the first printed book devoted entirely to mathematical games, this work entered the history of recreational mathematics. The games and puzzles (over 50 in total) always consist of problems to be solved using calculus and can be viewed as explanations of serious mathematical concepts in common and simple language. The subject matter revolves around various modes of transporting goods or people, card and number tricks, calculating weights, and other topics. The work was published in five editions (the last in 1959) and can be considered a forerunner to other collections of mathematical exercises for play that began to appear in print in increasing numbers up until our own time.

As previously stated, the philologist can provide little further commentary on the historical background of mathematical riddles. Franci's previously mentioned works (1981 and 2016) provide much deeper insights into the historical development of mathematical riddles in Europe.

---

2 Transl.: I believe that from the past chapter the appetite is annoyed by having written so much about commerce, so in this present chapter I intend to write about some delightful cases to soothe the mind.

4 Characteristics and Content of the Work

As mentioned, *Giuochi mathematici* is a work composed entirely of what we can define as mathematical riddles, even if labelling some portions of the text as riddles can sometimes be debatable.

In total there are 93 riddles in the treatise: 21 in the first part, 38 in the second, 27 in the third, and 7 in the fourth. The work becomes more organic with the introductory parts devoted to Giuliano de' Medici and his family, which present the obvious aim of winning the favour of the powerful family in whose service Piero da Filicaia's grandfather had already been.

A closer look at the structure of the work reveals that the first three parts depend heavily on Luca Pacioli's *De viribus quantitatis*, while the fourth part draws on Leon Battista Alberti's *Ludi matematici*.

Let us now examine the contents of the treatise's four parts or sections. The first section contains arithmetic riddles devoted to various methods of guessing a number. The riddles in the second part, as in the first, are about arithmetic, but there are also more elaborate games, magic squares, and even riddles with playing cards. In the *demostratio* (sic!) *quinta*, for example, Filicaia explains how to guess into which four parts a given sum was divided. Even for this section, the primary source, albeit modified, is Pacioli's treatise from a few years earlier. Filicaia becomes less schematic and repetitive in the later sections while remaining faithful to his sources. The third section consists of riddles and anecdotes (called cases) that do not require numerical calculations but rather reflection, such as the riddle about jealous husbands or the one about the division of wine from a barrel between two brothers (the main source is again Pacioli). Finally, the fourth section faithfully follows the first section of Leon Battista Alberti's treatise *Ludi matematici*, where the contents are related to geometry, for example, calculating the height of a tower or the depth of a well.

5 Luca Pacioli

After confirming Luca Pacioli's strong influence on the content of Filicaia's treatise, it seems appropriate to provide some information about this eminent individual. Luca Pacioli, the direct inspiration for and model of Filicaia's treatise, was the best known and most important mathematician of the late 15th and early 16th centuries. Rivers of ink have been spilled on this figure, particularly during his stay in Milan at the court of Ludovico Sforza, a friend of Leonardo da Vinci with whom he also shared his exile from Milan and wanderings after 1499. Pacioli's significance for economics and accounting has also been the subject of historical and linguistic studies (see Mattesini 2007; Ricci 1994; Sosnowski 2006). A brief biography of Pacioli was also published in the latter study, which is cited here with slight revisions (Sosnowski 2006, 63–69).

Friar Luca Bartolomeo Pacioli was born around 1446–1448 in Borgo Sansepolcro. The town is located on the outskirts of Tuscany. Some information about Pacioli's youth can be derived from his mentions in the *Summa* and testaments (Vianello 1896, 167–174) (Boncompagni 1879, 871–872) and from the documents recently found and studied by Elisabetta Ulivi (2009 and 2011). His father, Bartolomeo, must have been part of a family of some importance, if Pacioli's statement about his kinship with a certain Benedetto Baiardo, "mio stretto affine [...] capitano dei fanti, prima de lo re Alfonso in lo reame, poi de Sancta Chiesa al tempo di Nicola ...."[3] (Pacioli 1509, f. 23^{v}) is true.

Until his death Pacioli remained tied to his hometown where he was ordained as a Franciscan friar and to which he returned often, a guest of the convent of which he became commissioner in 1510. Despite some disagreements with the order, including the 1491 ban on teaching boys in his city (Yamey 1994, 15), his relations with the Franciscan brotherhood can be described as good. His election as commissioner and numerous passages in the *Summa* in which he expresses his gratitude to his superiors support this thesis (Vianello 1896, 35).

Pacioli spent his youth in Sansepolcro and especially in Venice, where he attended Domenico Bragadino's mathematics classes at the Rialto School and wrote his first treatise on arithmetic, intended for the children of the merchant Rompiasi, who was his host. His training as a mathematician was greatly influenced by his encounters and friendships with Piero della Francesca and Leon Battista Alberti (Vianello 1896, 15), in whom he inspired esteem and sympathy and of whom he was a sincere admirer.

In 1470–1471 Pacioli lived in Rome, then probably in Urbino (Ulivi 1994, 23), and in 1477 he became a teacher in Perugia with a salary of 30 and then 50 and 60 florins (Ulivi 1994, 23). Until 1480 he lived in Perugia where he taught mathematics (Ciocci 2017, 69) and in 1477–1478 he completed his treatise dedicated to the disciples of Perugia (Vat. lat. 3129) (Ulivi 1994, 23). In 1480 he stayed in Zara, then in Venice, Florence, where he met Piero Soderini, Rome, and Naples (Conterio 1994, 47). In those years his reputation as an exalted mathematician grew immensely, so much so that when he arrived in Naples, he had no difficulty making friends with the city's elites, who received him enthusiastically (Morison 1933, 3).

3 Transl.: my close kinsman [...] captain of the infantrymen, first of King Alfonso in the realm, then of Sancta Chiesa at the time of Nicola ....

Subsequent travels took him to Venice in 1494 to follow the publication of the *Summa*, to Milan in 1496 where he remained until 1499, called by Ludovico Sforza to teach mathematics at the Milanese *studium*. He became friends with Leonardo da Vinci, working with him and receiving his help. Leonardo made the drawings for Pacioli's work *De divina proportione* (1497); in return, Pacioli allegedly helped him with some calculations (it is believed that he calculated how much bronze would be needed to make the statue of Ludovico Sforza of Milan) (Jayawardene 1974, 269–272).

In 1499, he and Leonardo had to flee Milan because of the exile of their patron, Ludovico Sforza. Thereafter Pacioli taught in Florence, Bologna, then again in Perugia, Rome, and finally in 1508 returned to Venice to edit in person the edition of *Divina proportione* dedicated to Soderini. He lived an active life, between teaching at various universities and publishing works that increased his popularity and increased the number of invitations he received. In fact, in addition to the cities mentioned earlier, Pacioli's presence is attested in Padua, at the Montefeltro court in Urbino, and in Pisa. As he grew older, he perhaps limited his travels and teaching by returning permanently to the convent of Borgo Sansepolcro, of which he became commissioner in 1510. Once again, he could not resist the temptation to return to teaching mathematics: in 1514, we find him in Rome at the invitation of Pope Leo X. He was, therefore, also appreciated in Medici circles. The Roman period is the last record of Pacioli teaching. He died in 1517, between 15 April and 6 July (Ulivi 2009, 57).

Pacioli was, in the true sense of the expression, a Renaissance man in that he had acquired knowledge of different technical fields: commerce, military science, mathematics, medicine, art, music, and law. He believed along with his contemporaries in the need to link all these different disciplines and the special role of pure disciplines such as arithmetic and geometry. Above all, proportion, to which he devoted an entire treatise, was experienced by him and his contemporaries as a central issue of the varied disciplines such as architecture, military engineering, music, cartography, law, grammar, and painting (Yamey 1994, 12). Pacioli took a keen interest in art, in Neoplatonic philosophy, whose search for universal harmony also inspired mathematical research, and in the various aspects of daily life, as evidenced by the examples given in his works. He did not disdain the fame and glory that were not long in coming. Considered the most famous mathematician of his time, he was celebrated by artists, fought over by princes and universities. In 1508, in Venice, he delivered a lecture on Euclid, attended by all the leading figures then in the city. With a dose of vanity, in 1509, publishing a commentary on the Latin edition of the fifth book of Euclid's *Elements*, the result of this Venetian lecture, Pacioli enumerated 94 of the most significant (Nardi 1971, 69).

However, the work by Pacioli that interests us in the context of the mathematical games we have edited is *De viribus quantitatis*, a treatise that has remained in manuscript form, known through only one codex (Biblioteca Universitaria di Bologna ms. 250), published only relatively recently in Maria Garlaschi Peirani's transcription (Pacioli 1997).

De viribus quantitatis has several parts that are different from each other, as Montebelli (1998) notes. In the first one, devoted to the force of numbers, there are 80 *effecti* (Pacioli speaks of 81, but one of the *effecti*, namely No. 53, is missing). Such a "force of numbers" game is nothing more than a sleight of hand or parlour game. In the second part devoted to geometrical virtue there are 134 *capituli*, of which most (80) are devoted to the construction of geometrical figures with ruler and compasses (Montebelli 1998, 314) and others are curiosities, technical inventions, and physics-based conjuring games. In addition, there are other *capituli* that Montebelli calls topological (Montebelli 1998, 314) and a number of miscellaneous ones of a more general and sometimes literary nature, such as a poetic composition, various proverbs, documents, and merchants' proverbs (Montebelli 1998, 314).

The third part deals with "the natural power and virtue in writing". So, there are recipes for making invisible ink, for making good ink, for making hair dye, for making bombard powder—in short, a series of recipes that have to do with chemistry. However, these recipes are also designed to arouse the wonder of the company, so we ideally return to actual games—sleight of hand, company games that are meant to cheer and amaze. For example, we discuss how to cut fruit without cutting the rind, how to make an egg *walk* across the table, and similar things. After the third part, we find a collection of the heterogeneous material entitled *De problematibus et enigmatibus litteralibus* that consists of word games, riddles, and rebuses. Completing the treatise are riddles in the vernacular, amusing, curious, and even licentious, whose double meanings are justified by Pacioli at the beginning.[4]

As Montebelli (1998, 315) states, the content of Pacioli's work was not homogeneous, but at the same time it is a text that fits into the coeval tradition, that of "libri-zibaldone to which so many technical manuals, abacus books and workshop books then circulating belong".

The text is thus written for utilitarian, applied purposes, the theoretical topics having been addressed in the *Summa* published in 1494. The thread running through the entire treatise is the power of quantity (arithmetic and geometric), which is meant to arouse wonder and amazement. Montebelli speaks of "nat-

4 This part is entitled "problemata vulgari a solicitar ingegno e a solazzo".

ural magic that has its foundation in science" (1998, 317). In the end, *De viribus* is meant to entertain, sometimes indirectly, like when certain scientific paradoxes are presented that do not entertain per se, but induce reflection derived from amazement. Thus, they are books designed to intrigue the reader and inspire a benevolent attitude towards science.

Montebelli (1998, 319) notes that in many passages of *De viribus* he testifies to the prevalence of mathematical games and more in contemporary society. They were present in the abacus schools, the courts, and on the streets. The one that may be of most interest is the testimony about the courts where, according to Pacioli, c. 73^{v}–74^{r}, I, XXXIIII, a game is played that consists of two players alternately adding up numbers from 1 to 6, where to win one must reach 30 without going beyond it.

Grouping of mathematical games in *De viribus* (Montebelli 1998, 323–324):

1) guessing a thought number, integer, or fractional. 25 games. They are the *effecti*: 7–23, 26–31, 33, 45;
2) guess the parts into which a given number has been divided (2, 3, 4, 5 parts). They are 1–6, 39, and possibly 38, 41 and 44;
3) distribution of objects to people and then guessing who has what, possibly in what hand, etc. These are *effecti*: 35–37, 40, 42–43;
4) guess the thought object among the objects displayed on the table: 62, 64, 69, 74, 75;
5) arrange coins in successive piles, choose the pile, guess the contents of other piles: 76–78. In 79 the general rule "for finding the sum of primes in terms of an arithmetic progression" (Montebelli 1998, 324).

Then there are games of skill suitable as challenges—Pacioli explains how to win at these games. There are also games, quantified by Montebelli as 28 per cent (Montebelli 1998, 324), which are curious problems without the need to use mathematics, but only reasoning. They are: 46–61, 65–68, 70, 73.

Enzo Mattesini's analysis of Pacioli's language in *De viribus quantitatis* indicates that the choice of *volgare* was necessary given the text's audience, which did not know Latin (2010, 150). Simultaneously, he underscores that the language possesses a robust latinizing patina and warrants interpretation "as a composite expressive instrument, incorporating the typical elements of koiné languages, which are soon to succumb to the emergence and dissemination of a clearer and better assimilated Tuscanity. These languages, often rough and stunted due to their immaturity, are characteristic of many 15th-century writings in a technical-scientific register, among others".

Mattesini underscores how this text represents a significant phase in the evolution of Italian mathematical language, noting that Pacioli's use of terminology is extremely rich and, in some points, innovative in comparison to the previous works of Pacioli (2010, 166–167). This reflects the manuscript's purpose: to entertain and educate, thus addressing mostly laymen. Mattesini also emphasizes the widespread use of various words originating in dialects.

Mattesini's careful examination of textual inconsistencies suggests that transcription processes may have introduced some linguistic and terminological irregularities (2010, 143), but they also highlight the adaptive use of language to make complex ideas more accessible to a broader audience.

A comparison between the *De viribus* and Filicaia's *Giuochi* shows, on the one hand, the latter's great dependence on Pacioli (cf. below) and, on the other hand, the *Giuochi*'s greater compactness and concentration on the subject of games and riddles. From a mathematical point of view, neither the *De viribus* nor Filicaia's *Giuochi* are particularly novel, but they testify to a widespread culture at that time, the middle-class culture of the aristocratic and bourgeois strata. From a linguistic point of view, the treatises of Pacioli and Filicaia are very different. Pacioli's is a mixed language, and Filicaia's is pure Florentine, even though the author spent a long time in the same Borgo Sansepolcro from which Pacioli came.

Below is the scheme of comparison of contents between Filicaia and Pacioli prepared by Elisabetta Ulivi:

TABLE 1 Outline of a comparison between Filicaia and Pacioli: from Ulivi (2013, 256–257)

Piero da FILICAIA	Luca PACIOLI
Part One	
I	VII
II, III and XVII	X
IV	IX and XX
V	VIII
VI	XI
VII	XII
VIII	XIV
IX	XIII
X	XVI
XI	XV
XII	XVII
XIII	XVIII
XIV	XIX

TABLE 1 Outline of a comparison (*cont.*)

Piero da FILICAIA	**Luca PACIOLI**
Part One	
XV and XIX	XXI and XXVIII
XVI	XXVII
XVIII	XIX
XX	XXXI
XXI	XXXII
Part Two	
I	I
II	II
III	III
IV and XXXIV	IV
V	V
VI	VI
VII	XXII
VIII	XXIII
IX	XXIV
X	XXV
XI	XXXIII
XII	XXXIV
XIII	LXXXI, 115–118
XIV	119
XV	–
XVI	LVIII
XVII	LVI
XVIII	LXII
XIX	LXIV
XX	–
XXI	109
XXII	110
XXIII	XXXVIII and XXXIX
XXIV and XXV	XLI
XXVI	–
XXVII	XLII

TABLE 1 Outline of a comparison (*cont.*)

Piero da FILICAIA	Luca PACIOLI
Part Two	
XXVIII	XLIII
XXIX	46
XXX	XLIV
XXXI	XL
XXXII	XXXV
XXXIII	XXXVII
XXXV	XXXVI
XXXVI	LXIX
XXXVII	LXXII
XXXIX	LXXIV and LXXV
Part Three	
I	–
II	LXI
III	XLVII
IV	XLVIII
V	XLVI
VI	LXVII
VII	LXVIII
VIII	LXV
IX	72 and 73
X	77
XI	LXXIII
XII	82
XIII	LXVI
XIV	LIII
XV	LIV
XVI	113
XVII	89
XVIII	LXX
XIX	111
XX	112
XXI and XXII	LXXVI

TABLE 1 Outline of a comparison (*cont.*)

Piero da FILICAIA	Luca PACIOLI
Part Three	
XXIII	LXXVII and LXXVIII
XXIV–XXVI	LXXIX
XXVII	–

6 Leon Battista Alberti

Part Four of the *Giuochi mathematici* collects problems that, despite their undisputed mathematical character, are not riddles in the true sense. In the text, they bear the name *investigationes*.

As mentioned earlier, the direct source for this part of the treatise is Leon Battista Alberti's work *Ludi matematici* (also known as *Ludi rerum mathematicarum* or *Ex ludis rerum mathematicarum*). This is clear both from the text and from the iconography, which is almost identical. The illustrations on folios 166^{r}, 167^{r}, 169^{r}, 170^{v}, 172^{r}, 172^{v}, and 174^{r} of the manuscript Ital. Quart. 48 faithfully follow the illustrations in Alberti's treatise made more than half a century earlier (pre-1452). The *investigationes*, seven in number, are inspired by the text from the beginning of Alberti's work (these are problems 1 through 8, including 6a and 6b, while Alberti's third and fifth problems were skipped by Filicaia).[5] The first six *investigationes* concern height and width, while the seventh *investigatio* concerns depth.

At this point, the question naturally arises: Why did Filicaia decide to rework Alberti's treatise, which, incidentally, is only exploited in part (about a third)? This question is not easy to answer. The fourth part is not innovative, but it has in it something practical that perhaps in Filicaia's predictions might have pleased the treatise's dedicatee, the new lord of Florence, Giuliano di Lorenzo de' Medici, duke of Nemours. This, in addition to the title of *Ludi*, that is, "games", must have appealed to Filicaia so much that he incorporated, while reworking it, a part of Leon Battista Alberti's treatise. It is worth recalling the justifications Filicaia gives for his dependence on his predecessors, while at the same time asserting his own originality (f. 4^{r}):

5 This comparison of the content is based on Cecil Grayson's publication (Alberti 1973) and the article by Fabio Mercanti and Paola Landra (2007, 22–32).

> Potrebbe qua dire alcuno: Piero, sta fermo uno pocho; onde hai tu tracto questi tui secreti? Tu li hai rubbati. A che respondo brevemente che niuno in questo mondo è indovino né equale a omnipotente Idio che omni cosa sa et cognosce; et pertanto, se io ho racolto di più luoghi questi mia secreti et forze et forse parte rubbati et a questo e a quello, non sia in questa parte nessuno che me riphrenda [...] Et bene è virtù et ingenio mirabile et a suo proposito el saper rubbare et acomodare nelle cose virtuose.
>
> Someone might say here: Piero, stay still for a moment; where did you get these secrets of yours? You have stolen them. To which I respond briefly that no one in this world is a soothsayer or equal to omnipotent God who knows and recognizes everything; and therefore, if I have gathered these secrets and strengths of mine from many places and perhaps partly stolen from this and that, let no one rebuke me in this regard [...] And it is indeed a virtue and a remarkable talent, for one's own purpose, to know how to steal and adapt in virtuous matters.

Later, Filicaia quotes a proverb, as if to exonerate himself from accusations of not being innovative: "nihil dicitur *quod non* sit dictum prius". However, there also remains the entirely philological discussion of both the content and philology of the *Ludi albertiani* and the concrete witness of the *Ludi* that Filicaia had at his disposal. For the initial discussion, we refer to what has already been written by others (Alberti 1973, Geymonat 1980, D'Amore 2005, Mercanti and Landra 2007), and for the relationship between Alberti and Filicaia we limit ourselves to comparing the content, in the hope of being able to delve into the philological aspects as well at a later stage. The table comparing our treatise and Leon Battista Alberti's *Ludi* below allows us to verify the degree of correspondence between the two and the results of our comparison:

TABLE 2 Comparison of the contents of Filicaia's and Alberti's treatises

Piero da FILICAIA	Leon Battista ALBERTI
I (*investigatio* on measuring a tower)	1 (Measurement of the height of a tower, where the height of some point of it *is* known)
II (*investigatio* on the method of measuring a tower with the eye alone in a more difficult way)	2 (Measuring the height of a tower when the distance from its base is known)

TABLE 2 Comparison of the contents of Filicaia's and Alberti's treatises (*cont.*)

Piero da FILICAIA	Leon Battista ALBERTI
III (*investigatio* on another way of measuring similarly the height of a tower with the eye)	4 (Measuring the height of a tower with the help of a mirror)
IV (*investigatio* on measuring the distance of a river or similar width)	6a (Measuring the width of a river, first method)
V (*investigatio* one other way to measure a *distantia*)	6b (Measuring the width of a river, second method)
VI (*investigatio* other way to measure a tower)	7 (Measuring the height of a tower, which one *cannot* approach). Problem posed in a completely different way by Piero da Filicaia[6]
VII (*investigatio* to measure the depth of a well or the lake)	8 (Measuring the *depth* of a well).

7 Riddles Not Found Elsewhere

As noted, the majority of Filicaia's games are found in Pacioli's *De viribus quantitatis* (Parts I, II, III) or Leon Battista Alberti's *Ludi* (Part IV). However, there are some not recorded in those works.

These may be riddles absent from Pacioli's manuscript for philological reasons (several riddles reported in the index are missing in Pacioli) or different variants such that they are configured as autonomous riddles. In the former case, it is possible that Filicaia had access to Pacioli's text that contained such riddles. It must, however, be emphasized that, on the one hand, the "original" riddles are few compared to the total number and, on the other hand, that the setting is often very different in those clearly taken from predecessors (particularly in the arithmetic part).

The games not in the sources are:

6 In fact, it is not a reprise of Alberti's procedure, but a different method used to obtain the same result, namely the calculation of the height of a tower. It is possible that the witness of the *Ludi* that Piero da Filicaia had in his possession had a gap here, completed by the author using another procedure. Furlan (2006) also discusses inconsistencies at this point in the manuscript tradition of the *Ludi*.

- Part Two: XV, XX, XXVI
- Part Three: I and XXVII
- Part Four: VI

Moreover, some games that we know to be similar, probably present in the original *De viribus*, are absent in the only surviving copy of Pacioli's text. The index of *De viribus* allows us to establish the correspondence between Filicaia's and Pacioli's treatises. In the table of correspondence prepared by Ulivi (2013, 256–257), such games are indicated with Arabic numerals. Thus, we find in Pacioli's index, but not in the text, the games corresponding to these problems in our treatise: XIII (partially), XIV, XXI, XXII, and XXIX of the second part, and IX, X, XII, XVI, XVII, XIX, and XX of the third part:

TABLE 3 Problems present in Pacioli's index but absent in the surviving manuscript text

Filicaia	Pacioli	Game theme
Part Two		
XIII	LXXXI, f. 115–118	the game of *morra*, rule to always win
XIV	f. 119	the touch game with a trick that always allows you to win
XXI	f. 109	Uncover the card removed from the deck by looking at the deck quickly twice. The first time you do the calculation (addition) of all the numbers and figures. The difference between the result and the sum of the deck (312) reveals the value of the card. The second time you check the cards pay attention only to the missing figure or number. Cf. Bressanini and Toniato (2011, 109–110)
XXII	f. 110	Discover the card removed from the deck by looking at the deck quickly once. This consists of the combined calculation of numbers/figures and suits of the cards in the deck
XXIX	f. 46	Finding where a ring is hidden among members of a group
Part Three		
IX	ff. 72 and 73	A merchant sends his 4 hired hands to the market to sell 10, 30, 50, or 70 pieces of twill, respectively. They have to sell them with each getting the same amount. It resembles problem VIII, part 3, in which there are 3 henchmen.
X	f. 77	Rule for guessing how many pieces an apple is divided into
XII	f. 82	How to arrange a dance of eight women and seven men

TABLE 3 Problems present in Pacioli's index but absent in the surviving manuscript text (*cont.*)

Filicaia	Pacioli	Game theme
Part Three		
XVI	f. 113	Calculate how to arrange 10 people at a table without ever repeating the same arrangement. Told vividly through a story of 10 Florentine merchants eating at an osteria in Rome
XVII	f. 89	The case of nuns in a monastery who must be watched on site, with the ways in which they can be arranged so that they can be monitored, and the ways in which they can be arranged so that they are not monitored
XIX	f. 111	Grains of wheat are doubled every day for a month. Gives an example of geometric progression: 1, 2, 4, 8, 16, 32, 64, 128, 256, 512, 1024 … etc. How to calculate the sum of this progression in a simplified way.
XX	f. 112	Grains of wheat are doubled (after adding up the previous numbers). Example: 1, 2, 6, 18, 54, 162, 486, 1458 … etc. How to calculate in a simplified way the sum of such a progression

CHAPTER 3

Linguistic Analysis of the Treaty

1 Aspects of the Language of the Treatise

We can assume that the two extant manuscripts were reviewed directly by the author and represent the original language. Though minor, there are some differences between them. We use the manuscript Ital. Quart. 48 and quote the forms from that document in our examination of the language. The findings of a linguistic analysis of Magliabechiano XI, 15 in the National Library of Florence would not be radically different. We try to identify any more substantial divergences.

The treatise's language can be characterized as fluid and condensed. A manuscript from the first half of the 16th century does include some variations, but they are few. By contrasting the work with other writings and accounts of Florentine culture from the 15th and 16th centuries, we were able to analyse Filicaia's language. The following studies on Florentine served as our main points of reference:

- Manni, Paola, 'Ricerche sui tratti fonetici e morfologici del fiorentino quattrocentesco', *Studi di Grammatica Italiana*, VIII (1979), 115–171.
- Manni, Paola, 'Dal toscano all'italiano letterario', in *Storia della lingua italiana*, ed. by Luca Serianni, Pietro Trifone (Torino: Einaudi, 1993), III, 321–342.
- Frosini, Giovanna, 'Firenze', in *Città italiane, storie di lingue e culture*, ed. by Pietro Trifone (Roma: Carocci, 2015), 203–224; 216–221.
- Frosini, Giovanna, 'Lingua', in *Enciclopedia Machiavelliana*, 3 Vols (Roma: Istituto della Enciclopedia Italiana Treccani, 2014), II, 720–732.
- Franco, Matteo, *Lettere*, ed. by Giovanna Frosini (Firenze: Accademia della Crusca, 1990).

1.1 *Selected Orthographic Features*

Strong Latinizing tendencies are particularly apparent at the orthographic level, closely resembling the preferences and approaches of Tuscan writers of the late 15th and early 16th centuries (we find similar features, e.g., in the famous letter to Francesco Vettori written by Machiavelli on 10 December 1513). Several etymological solutions are found, such as **ti** followed by a vowel: *scientia* (1^r ff), *conditione* (1^v); preservation of the etymological **h**: *mathematica* (1^v), *habbi* (1^r), *havere* (1^r), *homo* (1^v), *hoggi* (1^v), *havendo* (2^r); preservation of the consonantic cluster **ct**: *proficto* (1^r), *perfectione* (1^r), *sancta* (1^v), *sopradicte* (1^v),

 | DOI:10.1163/9789004727984_004

tucti (2^{r}), *riducti* (2^{r}), *sanctissima* (1^{v}) (cluster *-ct-* with 1,441 occurrences in the whole text); preservation of the consonantic cluster **pt**: *acceptare* (3^{r}), *scripto* (3^{r}), *captivi* (24^{r}) with numerous occurrences (129); preservation of the consonantic cluster **ps**: *epso* (1^{v}, 3^{v}), *relapsata* (124^{r}); presence of the etymological **x**: *exaltare* (1^{r}), *expositioni* (2^{r}), *exemplo* (4^{r}). We have also the consonantic cluster **ph** preserved in some words as *philosopho* (1^{r}), *triumpho* (63^{v}), *dalphini* (120^{v}), and a number of times **h** is also present in words where there is no **h** in Latin, as in the verbs *rephrendere* and *comphrendere* (2^{r}, 7^{r} ff.) or in forms like *sapienthia* (44^{r}), *thrare, thrarai* (97^{r}).

We can observe the oscillation of the velar occlusive **ch** against **c** in front of a back vowel that heralds the modern phase of the Florentine language (Blasco Ferrer 2015, 85) and is typical of that period in Florence. For example, *amicho*/*amico*: "Poniamo che *lo amicho* habbi preso 8 et 3/4, doppiato fa 17 et ½, postovi in su 17 ½ 5 fa 22 et ½, multiplicato 22 ½ per 5 fa 112" (11^{v}) and "Et se *l'amico* pigliassi 5 et 1/2, doppiato fa 11, postovi su 5 fa 16, multiplicato per 5 fa 80, postovi su 10 fa 90, multiplicato per 10 fa 900, tractone 350 resta 550". We also reckon *eccho*/*ecco*: "***Eccho*** lo exemplo materiale: sia la distantia che è fra voi cioè fra l'occhio vostro et il piè della torre sia braccia 400" (167^{v}). "*Behold* that the second time you did not break (8^{v})." Other examples are: *long*/*longho*, or the verbs *venga*/*vengha* and *ponga*/*pongha*.

With great regularity, the nasal palatal is written **gn**: *digni* (2^{r}), *digno* (2^{v}), *cognitione* (3^{r}). Sporadically, the nasal palatal is spelled **ngn**: *ingnoranti* (105^{r}), *ingnorante* (137^{v}), *ingnoranti* (138^{r}), *ingnorante* (139^{v}).

In some words, the consonantic cluster **pt** can be found even if it is not etymological. The best example is *ciptà*: "Advenne che, forte fortuna, capitò nella *ciptà* di Milano uno architectore fiorentino." In another place, we find *presunptuoso*: "Ho voluto, Magnifico Giuliano, questa brieve faccetia recitare perché forse sarà qualche uno più ***presunptuoso***" (46^{v}).

The alveolar affricate is represented by **z**: *mezo* (5^{r}).

The use of capital letters is uncertain and unstable, a feature also documented in Franco's letters (Franco 1990, 164).

In Filicaia, as indeed in Matteo Franco's letters (Franco 1990, 165), the bars or colon on the side can circumscribe the verbal voice is (/e/). The punctuation is in fact entrusted to a few graphematic signs, not incorporating the innovations, later widespread, introduced by the edition of *De Aetna* (1496) and subsequent Bembian aldines. A half-height dot stands for a weak break but is often placed where we would expect no break. Some regularity (though not absolute) is found in the signalling of the voice of the third-person singular of the verb to *be* (form *.e.*), which is signalled with dots on either side of the word. The same dots on the side are used to highlight both Roman and Arabic numerals.

Examples:

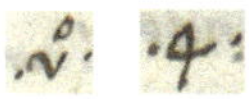

FIGURES 1–2 Roman and Arabic numerals with dots on both sides

1.2 *Selected Phonetic Features*

There is an oscillation between diphthongate and non-diphthongate forms at the phonetic level in the otherwise fairly regular text: *giochi* (1^{r}, 4^{r}), *novo* (9^{v}), *vole* (35^{r}), but also *giuochi* (4^{v}), *giuocho* (6^{v}), *may* (4^{v}). It is worth noting that the two witnesses (K and F) differ in their diphthongate and non-diphthongate forms. Sometimes the diphthongate form in K corresponds to non-diphthongate in F and vice versa. Overall, the diphthong appears to have been reduced in line with what was going on in late 15th- and early 16th-century Florence. In comparison to the letters of Matteo Franco studied by Frosini, Filicaia has a greater reduction of diphthong. The reduction of the diphthong in *trovare* (rather than *truovare*) is particularly significant because normally the diphthongate form is tenaciously preserved and "will turn out to be a minority only beginning in the second or third quarter of the sixteenth century" (Franco 1990, 167). However, *brieve* (with fewer occurrences than *breve*) is present, which was more often monophtongated in early 16th-century Florentine scripts.

We also note the exchange of the cluster **ng** with the palatal nasal, also typical of Florentine. We find, for example, "Onde tucti stupiti quivi non fu nessuno che potessi questo effecto mandare ad executione, ma guardando l'uno l'altro si *ristrignevono* nelle spalle" (46^{r}), but also *giugne* (72^{v}) for *giunge* and similar others. Moreover, the **gn** is found in the verb *porre* and *venire*. We read, "***Pognamo*** *che* l'amico pigliassi"; elsewhere, "attendiamo all'opera nostra et *vegnamo* alla prima dimostratione di questa seconda parte" (47^{r}).

The closing of /**e**/ in protonia is also a notable feature. Several examples are observed throughout the treatise, among which the *dimandare* form expressed in different verb tenses stands out: e.g., "**et tu *dimanderai*** che differentia sia dalla tua somma alla loro" (109^{v}) and other, numerous occurrences (49).

Labialized forms like *domandare* are also present with numerous occurrences (29). Thus, in the same text we have the oscillation between *dimandare*/*domandare* with the first form slightly more frequent.

Another type of oscillation concerns /**o**/ and /**u**/ in protonia, where *u* is often retained (in borrowings from Latin in particular), e.g., *facultà* (2^{r}), *multiplicare* (38^{v}), *subtractione* (3^{v}) as indeed happens in Franco's letters (Franco 1990, 171).

Anaphonesis is the articulatory raising of vowels /**e**/, /**o**/ before the velar nasal or the palatal (they become /**i**/ and /**u**/ respectively). It is a typically Florentine phenomenon, one of the phonetic traits that most demonstrates the Florentinity at the base of Italian. The trait is present in our treatise in forms that are continued in modern Italian and with some instances of extension from modern usage. For example, in the treatise we find both *digno*, and *degno*: "anchor dire questa ***digna*** **scientia** havere in sé qualche parte di divinità" (1ʳ), but also "Tamen lo ritrovai et parmi *degno di* essere commemorato nelli mathematici ludi et è questo" (136ᵛ). Also recorded are *signare*, "hai a pigliare la seconda figura comme vedi qui *signato*" (108ᵛ), and *segnare*: "potete fare con uno filo piombinato facendolo pender innanzi ad voi et *segnare* in epso le mire vostre" (166ᵛ).

Phonosyntactic doubling is graphically signalled in only a few cases: in the case of the preposition *a* with the pronoun *he* (*allui* 33ʳ), in the case of the preposition *a* with the verb *fare* (*affar* 33ʳ).

An protonic instead of *en*, the traditional Florentine usage (Frosini 2014, 724), is only partially present. Next to a few instances of *sanza* (4), we find an overwhelming majority of *senza* (52).

Some more popular Florentine features, such as **sti** per **ski**, widely found in 16th-century writings, e.g., in Machiavelli (Frosini 2014, 723) or **r+consonant** replacing **l+consonant** (*fragello* instead of *flagello*), are absent in Filicaia (ibid.).

1.3 *Selected Morphological Features*

Examining the morphology leads to the conclusion that the text is very compact and mostly conforms to the linguistic solutions common in late 15th- and early 16th-century Florence. There are few uncertainties; the various forms, even if some oscillation is present, are firm and unambiguously interpretable. Filicaia's language, in its morphological solutions, is undoubtedly Florentine, as if the author had never left Florence. There is no obvious trace of his stay in Borgo.

The following are the various morphological phenomena that testify to the Florentinity of the text of Filicaia's treatise.

1.3.1 Determinative Article

The determinative article shows strong fluctuations, a fact not surprising for that era. The competing forms are **lo**/**il**/**el**, the latter typical of the Florentine argent. Here are some examples: "*Il che* inteso *el* Magnifico duca di tale cosa sbigottito più et più anni la lasciò inperfecta" and "***el*** **duca** comme desideroso de intendere et vedere li homini virtuosi un giorno lo fece ad sé venire," "apuncto *el* numero che lui prese comme se *lo* amico tuo havessi preso" (62ʳ); (*el* numero 1ʳ, *el* partire 3ᵛ, el preterirla 3ᵛ, *el* rotto 5ᵛ).

Between the 15th and 16th centuries, the morpheme **el**, in terms of frequency of use seems to be the prerogative of a popular strand with little evidence, e.g., in the *Stanze del Poliziano* (Blasco Ferrer 2015, 95). In the case of Filicaia, it oscillates between **el** of the Florentine argent and the cultured Florentine norm of **il** (with **el** being the majority). Regarding plural articles we find the widespread use of **li**: "*li* meriti del Magnifico Lorenzo inverso di lui" (2ᵛ). Less frequent is **i** and sporadic **gli** "tucti *gli* altri" (141ᵛ), "among others" (176ʳ). Significant is the form **ei** (e') which occurs in a limited number of cases. It is not a completely unknown form in 16th-century Florentine texts (present, e.g., in Jacopo Nardi), but uncommon: placed above *ei* dua quadrati (13ʳ), cavi fora *ei* grossi (56ᵛ), tucti 3 *e'* numeri (54ᵛ).

1.3.2 Possessive

The possessive system can be defined as typical of the period and diatopic variety. We encounter the indeclinable forms: **mia**, **sua**, **tua** in the plural. We read, e.g.: "dalli *mia antecessori et* antiqui" (44ᵛ), but also "e *sua* architectori erano consueti" (45ᵛ), "*mia* anni" (1ᵛ), "*sua* primi fundamenti" (3ʳ), "questi *mia* secreti" (4ʳ). This is the 15th-century development present in Florentine (Manni 1993, 322).

1.3.3 Personal Pronouns

The subject pronoun is now exclusively **lui**, which testifies to the completion of this change in Florentine. **Lui** has 446 occurrences as both subject pronoun and oblique pronoun. **Egli** is present exclusively in the expression **egli è** with the meaning *there is*, while it does not appear as a subject pronoun. "I know that *he is* in the world of *omni sorte de homini*" (30ʳ). The feminine subject pronoun **lei** is present only once and refers to the inanimate subject: "A questo non rispondo perché l'opera risponderà lei" (24ʳ). For the plural, only the pronoun **loro** is used.

1.3.4 Numerals

There are a large number of numerals, as is to be expected in a text with many calculations. Significant are the forms of the numeral **due** (two), which varies considerably, as is normal in contemporary Florentine texts (Frosini 2014, 727). The most frequent is **dua** (103 occurrences), **due** is also present (33 occurrences), **dui** (2 occurrences), and **duoi** (1 occurrence) are marginal. In Latin inserts, we also find **duo**.

1.3.5 Other Phenomena

The plural formation includes the feminine forms of the third declension in **-e** (*le parte*) and the plural forms of the feminine adjective in **-e**: *cose sottile* (3ᵛ), *cose dilectevole* (4ᵛ), *le quale cose* (5ʳ).

Also present is the type **in nel** per **nel**, which is described as being of western origin (Manni 1979, 168), and consistently found in Machiavelli as well. On the stable presence of *in sul* in Machiavelli's prose, see Frosini (2014, 727). The preposition **sul** also appears in the strengthened version **in sul** also in *Giuochi*: 12ᵛ, 93ʳ, 114ʳ, 148ʳ, and 25 other occurrences.

1.3.6 Verbs

In the present tense, as can almost be taken for granted in a Florentine text, we find the first-person plural of the verbs **-iamo** instead of **-imo**, **-emo**, **-amo**: "Per questa tucta preallegata ragione *possiamo* et *dobbiamo* dire veramente questa scientia esser divina et in omni genere perfecta" (1ᵛ).

Analogical ending *-ono* in the remote past and conditional tense.

Examples of this analogical ending are not lacking in the text (see also infra the imperfect). Remote: "vari architectori dilectandosi intendere ***vennono*** a vedere el modo e l'ordine" and again "*mostrorono* alla excelentia del signor Duca", "Tucti una voce ***dissono***", "in modo che tucti i christiani ***rimasono***" (78ʳ), "le promesse grande che lui mi faceva me ***spinsono*** ad acceptare" (120ʳ). Conditional: "mancho assai *harebbono* loro tale ponte fabbricato" and "in *quale modo* anchora loro *saprebbono* porre l'uovo".

Present Subjunctive

The third-person singular and third-person plural of the present subjunctive in **-i** **-ino** is due, as Frosini argued (Frosini 2014, 727; Manni 1979, 156–159), to analogical drives in verbal paradigms. These forms are widely attested in Machiavellian autographs and, not surprisingly, are present in the text of our coeval and Florentine treatise: *habbi, vadi, facci, vadino, faccino, habbino*. As in Machiavelli, an oscillation is noticeable here because the **-a** forms are present (to a lesser extent): *faccia, vada, habbia, habbiano* (*facciano* and *vadano* are absent).

Future Tense

In the future forms, we observe a wide extension of the vibrant intense: "et sebene fussino 100 o 1000 homini sempre la *troverrai*" (99ʳ); "la sua alteza *entrerrà* nella distantia" (167ᵛ), and many other examples.

In the future, we note the presence of the simplified form **arai** (**harai**) accepted in Florence after 1350 (Manni 1979, 141), spread from Western Tuscan dialects. This form is often used when teaching precisely mathematical games: "Et in tucto *harai* 20 monti" (126^{v}), "in sé *arai* 2 1/2 in sé fa 6 1/4 et 4 in sé fa 16" (13^{v}). *Harò* per *avrò* (4^{r}), *harai* (6^{r}) are traits that can easily be classified as 15th-century Florentine forms.

Imperfect

The imperfect in the second-person singular is analogical, as imposed in Tuscany in the 15th century: [tu] *havevi* (6^{v}). The return to the 14th-century usage conditioned the use in the Italian literary language of the etymological form [tu] *haveva/aveva* for many centuries, while in Florence people continued to say [tu] *avevi*. The first person is also analogical in **-o**: *pigliavo* (119^{r}), *sapevo* (122^{r}), *volevo* (123^{r}). The ending in **-a** is used exclusively for the third person.

We also note *-ono* instead of *-ano* in the imperfect, a trait believed to be a 15th-century evolution in Florentine (Manni 1993, 322), e.g., *pensavono* (125^{r}) and *raccomandavono* (146^{r}).

Passato remoto

The type *missi* per *misi* (once) is present: et così li missi d'acordo (139^{r}). For the analogical desinences **-ono** cf. supra.

Conditional Forms

The conditional of the third-person plural in **-ebbono**: *varrebbono* (2^{v}) is prevalent, but here, too, we find the oscillation—typical from the late 15th century onward—with non-Florentine forms such as *saria*, *faria*, interpreted by Manni (1993, 322) as a southern influence. In the case of *saria* (for *sarebbe*) this is a more frequent form than *sarebbe*; *faria*, on the other hand, is a clear minority form in our text.

Another interesting phenomenon is the fluctuation in the output for the third-person plural of the present conditional between **-ono** and **-ero**. We have both forms, such as *conphrenderebbono*, *potrebbono* and also *farebbero*, *potrebbero*.

Imperfect Subjunctive

The forms of the third person of the imperfect subjunctive with the ending in **-i** instead of **-e** are thought to be typical for the fifteenth-century Florentine vernacular: "Poniamo l'amico *pigliassi* 8, factone 2 parte, 3 et 5 quadra ciaschuno harai 9 et 25" (14^{r}), or "et così tornerà sempre quando l'amico *pensassi* uno numero rotto solo et partissilo in più parte rotto" (14^{v}). In our text, we have alternating forms between those with **-i** and with **-e** (*havessi* and *havesse*) with the former being more frequent. As for the plural, the late fifteenth Florentine form type *havessino* for *havessero* is almost exclusive: "se li antiqui doctori di medicina non le *havessino* cognosciute et datone notitia?" (2^v), "non mi pareva *havessino* potentia di potere epsa submergere" (119^v). Only once do we have *volessero* (129^v) (vs. 4 *volessino*).

The Verb to Be

The present subjunctive is in line with contemporary Florentine prose. If we take Machiavelli's *Prince* as a point of reference, we will find exactly the same forms: **sia** for the singular and **sieno** for the plural and sporadically **siano** (one occurrence in Machiavelli and one also in Filicaia). Thus, we can see a difference between the singular and the plural. **Sieno**, the form that continues the literary language of the 14th century, is firm while **sie** is absent. Here are various examples:

"tucti sono variati nel conspecto delle persone parranno varii et sono perché, benché *sieno* in una medesima regola fundati" (11^r) "ti piglia dua moltiplichationi l'uno paro et l'altro caffo et *sieno* di che qualità o quantità si voglino pure che non *sieno* rotto" (104^v). The form **siano** (1 occurrence): "purché *siano* differentiate et diverse perché non potria l'occhio tanto apertamente judicare" (112^r). **Sia** (with 233 occurrences) e.g., "voi direte bene che io *sia* valente homo?" (46^r), "uno che *sia* a lato al secondo" (117^r).

Also regarding the verb **essere** (to be), there is the imperfect subjunctive in the first, second, and third person, **fussi**, with the tonic **u** that is found from the very beginning in the Sienese, Pisan, and Lucchese and interferes in Florentine after the first decades of the 14th century (Blasco Ferrer 2015, 100). **Fussi** and **fussino** are generalized forms, whereas **fossi** and **fossino** are not found. Some examples would be: "et che con mancho spendio et più facilità che a lloro *fussi* possibile li dessino disegno" "perché lo amico possa discernere quale lui vole pigliare perché *se fussino* tucti in uno conio con difficultà potrà Vedere" (81^r and 81^v).

Also present are the forms *suto* (3^r) for *stato*, incidentally frequent in Florentine writers of the period, and, equally, *fia* (7^v, 8^r).

1.4 *Selected Lexical Features*

A considerable repetition of words in the text must be noted before considering the lexicon. This should not come as a surprise considering that at least the first and second parts involve word-based maths games with several calculations.

The third section, which has more logic games than maths games, and the fourth section, which is devoted to geometry, are lexically distinct but nonetheless repetitive in themselves. The introductory passages to each portion of the book, when the author displays his literary ambitions, are where the text deviates significantly and has more linguistic complexity. These sections are intended to praise the Medici family and, in particular, the dedicatee. As a result, they contain poetic visions with a certain fervour and most likely reflect the author's efforts to display his literary skills in order to gain acceptance into Giuliano de' Medici's circle of devoted followers.

The well-known measure of lexical richness, Giraud's index (not lemmatized), is 14.93, which corresponds to a highly repetitive text. The average for 20th-century novels is 28.30 (calculation based on novels in the *Primo Tesoro della Lingua Italiana del Novecento* PTLIN), Giraud's index for Bembo's *Gli Asolani* is 22.48.

In several instances, the repetitiveness and formulaic nature of Giuochi's text make it impossible to tell whether particular morphological forms would be present or not. Alternatively, because it must be considered in the context of formularity, the absence of some forms or words is not always important in and of itself. For example, due to the very nature of Filicaia's considerations, a number of verbal forms are missing.

One should also take into account the terminological element and its relationship with the mathematical sources of the treatise described above (Pacioli's *De viribus quantitates* and Alberti's *Ludi mathematici*; see also below). After a careful analysis is carried out by Palmarini and Sosnowski (2019; 2020), we conclude that most of the mathematical terms used by Filicaia corresponded to those of his sources (Pacioli's *De viribus quantitatis* for the arithmetic part and Alberti's *Ludi* for geometry). There is, however, an internal consistency in the use of terminologies. This helps to give our treatise the appearance of homogeneity.

At this point, we will only mention a few words in the treatise that are more marked in the diatopic sense, being reported by Manni (1979) as typical of that variety of Florentine: *drento* instead of *in* (only once in manuscript F); *drieto* instead of *dietro*; *fuora* and *fora* instead of *fuori*.

2 Specialized Language in the Treatise

Below, we go on to analyse the presence of specialized language used by Filicaia in the treatise, which will first be compared with that of Luca Pacioli and subsequently with that of Leon Battista Alberti.

2.1 *Comparison of the Language of Piero da Filicaia with the Language of Luca Pacioli*

The comparison with Pacioli's language must encompass at least two treatises: *Ad discipulos perusinos* and *De viribus quantitatis*. The second is necessary for its content and for chronological reasons, while the first is necessary for philological and linguistic reasons. *De viribus quantitates* contained in ms. 250 of the University Library of Bologna (the only extant copy), does indeed contain riddles directly comparable to the Filicaia text and was a direct source of inspiration for our author, but it is a text that has been altered linguistically by the copyist (Pacioli 1997, VIII). We can, however, access Pacioli's genuine language through the text of the treatise written in 1478 (treatise *Ad discipulos perusinos*), contained in the Vatican Latin codex 3129, which is one of the two known Paciolian autographs (Bressanini and Toniato 2011, 7). The Perugian mathematical treatise also contains the description of mathematical games recently published by Bressanini and Toniato in a volume that combines philological precision with historical and mathematical insight. A look at Paciolian language points out numerous similarities and differences at the phonetic and morphological levels between Fra Luca and our Piero da Filicaia.

The limitation of the diphthong, already present in Filicaia, is an even more conspicuous tendency in the Paciolian language, moreover in accordance with the characteristics of the language of Borgo Sansepolcro. While in Filicaia we find anaphonetic forms (*giugnedolo* 5v, *apuncto* 171v), in Pacioli they are absent (*agionga, ponto, longi*). The first-person plural in **-iamo**, naturally present in the Florentine Filicaia, does not appear in Paciolian writing where we find the forms: *ponamo, metamo* and not *poniamo, mettiamo*. As Enzo Mattesini (2007, 53–62) reminds us, other anti-Florentine features of the Paciolian script are: the preservation of the **-ar-** intertonic in futures and conditionals (e.g., *andarà, avarà, avanzarà, sumarai*) as well as the extension of **-ar** to cases where there should have been **-er** as a base; the disempowerment of some long consonants of northern origin (e.g., metici, tute) that alternates with the majority of non-de-emphasized forms; at the morphological level, the forms of the conditional that continue infinitive + HABEBAM (e.g., *andaria, bisognaria*); the extension of the ending of the third-person plural of the present indicative of verbs of the first conjugation **-ano** to the other conjugations (Mattesini 2007, 61). We can thus see that, at crucial points, Filicaia adheres to the contemporary Florentine norm, rather than following the local variety of the place where he resided, namely Borgo Sansepolcro, a language that we instead find represented in Luca Pacioli's writings (Sosnowski 2014).

2.2 *Mathematical Terminology in Filicaia's Treatise*

The following table summarizes the mathematical terminology used by Filicaia in his manuscript and that found in Pacioli's *De viribus quantitatis.*

TABLE 4 Comparison of Filicaia's mathematical language with Pacioli's

Luca Pacioli	Piero da Filicaia
aboco, ***abaco*** "arithmetic practice" *De viribus, effecto xxxvii* El dare 12 a uno, et 24 a un altro, et 36 a l'altro, se po facilmente per li idioti et quelle che non sanno **aboco** nè lor tra- vagliamenti, farla con fave, o vero quartaruoli, o altra moneta in questo modo	**abbacho,** ***abaco*** "arithmetic practice" *Demostratio xvii, parte ii* [80^{r}] Così *quando* havessi a essere 18, 20,15, et 12, potrai aseptare el tuo **abbacho** et veder*e* in ch*e* luogo tu poi stare p*er* ess*ere* francho et p*er* 7, et p*er* 9, o veram*en*te tucti a dua nu*mer*i no*n* noce
asettare "to place, to arrange neatly" *De viribus, capitolo lxii* La qual cosa farai in questo modo, videlicet: prima o vero poi, tocho che lui ara o ver pensata, et tu ordinatamente le **asettarai** in un cerchio et tacitamente tirai el numero de tutte ditte monete, o ver persone, o ver cose etc.	**aseptare** "to place, to arrange neatly" *Demostratio xvii, parte ii* [80^{r}]veder*e* in ch*e* luogo tu poi stare p*er* ess*ere* francho et p*er* 7, et p*er* 9, o veram*en*te tucti a dua nu*mer*i no*n* noce. Et sirà raro ch*e* se tu harai a m*en*te e' luoghi franchi ch*e* no*n* ti sappia o possa **aseptare.** Et qui si mostra qua*n*to vaglia [80^{v}] ess*ere* bene exp*er*to i*n* matematicha
acozare "to sum up" *De viribus, capitolo cxxix* Quando te fosse ditto: "**acozame** 3 tagli de coltello insiemi a modo triangulo, che si tengano insiemi levati e che, postovi su un gran vaso pieno, lo tenghino"	**acozare** "to sum up" *Demostratio vi, parte i* [13^{v}] **Acoza** insiemi fa 2/16, m*oltiplic*ato l'una nell'alt*r*a fa 1/16, adoppiato fa 2/16, posto sopra el quadro delle pa*r*te in se fa 4/16, ch*e* è ¼, la cui radice è ½ et questa regola mai fallirà et è bellissima.
avanzare "to remain as a remnant in subtraction and division" *Summa* [121^{v}, 4] **Avança** el quadrato della R(adice)	**avanzare** "to remain as a remnant in subtraction and division" *Demostratio vii, parte ii* [62^{r}] et tu tieni alle mani 70 et *quando* lui lo pa*r*te per 5 p*er* omni unità ch*e* li **ava*n*za**, et tu tieni alle mani 21

TABLE 4 Comparison of Filicaia's mathematical language with Pacioli's (*cont.*)

Luca Pacioli	Piero da Filicaia
caffo "odd" *De viribus, effecto IV* Et per più ampia libertà dirai a loro medesimi che gli batizino, ma che non metino el medesimo numero a più che una carta, o paro o **caffo** che sia, perché, nel tuo operare per le date regole, retroveresti el numero de cadauna carta che fra loro fosse distribuita	**caffo** "odd" *Demostratio XIV, parte II* [76^{v}] Ma s'el compagno volessi ess*e*re toccho lui, allora tu dirai ch*e* alzando puncto pari vuoi ch*e* le si giungha 5 o 7 o 9 o altro pu*n*to ch*e* sia **caffo**, purché el nu*mer*o dive*n*ti **caffo** p*er*ch*é* el nu*mer*o **caffo** sempre te*r*mina in lui. Alzando **caffo** no*n* bisogna porvi legge
chavare (cavare) "to subtract arithmetically" *Ad discipulos* [1.1] Egli è uno ch'à preso un numero in la sua mente e dici "fame far quello che tu vuoi" purché non me ne faci **chavar** alcuno	**chavare** "to subtract arithmetically" *Demostratio II, parte II* [51^{r}] Hora dirai che **chavino** la lor somma, cioè 99, della tua, 154, resterà 55.
cubare "to elevate to the cube" *De viribus, effecto XII* et **cubata** ditta parte fa $^{2197}/_{27}$. Et l'altra parte quadrata fa $^{361}/_{9}$, et triplata fa $^{1083}/_{9}$, che multipli-cato via l'altra parte fa $^{14079}/_{27}$, et cubata farà $^{6859}/_{27}$	**cubare** "to elevate to the cube" *Demostratio VII, parte I* [13^{v}] m*oltipli*chi poi via la p*ri*ma pa*r*te et il quadrato della [14^{r}] p*rima* i*n* 3, m*oltipli*chi nella *seconda* et di poi **cuba** omni una di dicte pa*r*te di per se et tucto acozerai i*n*siemi. Et dicati la som*m*a tucta dela quale tu piglierai la sua radice cuba et quello fia el n*umero* p*reso*
figura "digit, number between 0 and 9" *Ad discipulos* [19.2] E fatto questo domandarai la summa e lui te dirà i nove letere commo sonno 9cose, e tu vedarai quelle **figure**, e recognoscerai quelle che tu desti e a chi, zoè el 2, el 3, el 4 ech.	**figura** "digit, number between 0 and 9" *Demostratio XX, parte I* [37^{r}] Hora per questa seco*n*da m*oltipli*catione tu sai che l'ultima **figura** dello tuo amico è 2 p*er*ch*é* 6 m*oltipli*cato p*er* 2 p*ro*durrà p*er* sua ultima **figura** 2, ch*e* 2 via 6 fa 12
giognere "to add, to add arithmetically to a given quantity" *Ad discipulos* [3.2] Dirai che c'**agionga** la ½ di quello ch'egli à preso e poi anche dirai che pigli la ½ di tuta la summa che lui à e poi **la giogna** sopra a quello ch'el à.	**giugnere** "to add, to add arithmetically to a given quantity" *Demostratio IX, parte II* [67^{r}] A questo **giugni**, uno fa 721. Et questo farà el bisogno. Et da te col tuo ingegnio non potrai trovare infiniti

TABLE 4 Comparison of Filicaia's mathematical language with Pacioli's (*cont.*)

Luca Pacioli	Piero da Filicaia
multiplicare "to multiply" *De viribus, effecto xxxii* de doi numeri che, **multiplicato** l'uno in l'altro, sempre farà la summa del producto le figure che voli	**multiplicare** "to multiply" *Demostratio iv, parte i* [10^{v}] La prima volta **multiplica** 15 p*er* 3 fa 45, pa*r*ti p*er* 2 ne viene 22 *et* ½ che harai rotto. La *second*a volta dirai lo facci intero, fa 23 e tu p*er* questa *second*a volta tieni alle mani 2
partire "to divide" *Ad discipulos* [3.7.2] Fa 58 e 1/2, e questo **parti** in 9 ne ven 6/2	**partire** "to divide" *Demostratio iv, parte ii* [53^{v}] dirai che cavino tucta la lor somma, et quello resta tucto da te **partirai** per 2 mancho che il numero per cui moltiplichasti el numero diviso.
radice cuba "given a certain number, a number which, raised to the third power, results in the given number, cube root" *Summa* [115^{v}, 13–14] Per Radice cuba se intende n(umero) in se moltiplicato el quale de tal producto esso fia dicto esere. Radice. come 4 e ditto R(adice) de 16	**radice cuba** "given a certain number, a number which, raised to the third power, results in the given number, cube root" *Demostratio vii, parte i* [14^{r}] dicati la som*m*a tucta dela quale tu piglierai la sua **radice cuba** et quello fia el n*umer*o p*r*eso
aredopiare "to double, to multiply by two" (also **dopiare**, **redopiare**) *Ad discipulos* [4.2] Di ch'el **aredoppi** quante volte che tu voli e tante volte quante lui **redoppia** el suo numero **redopierai** uno	**radoppiare** "to double, multiply by two" *Demostratio iii, parte ii* [52^{r}] cioè per 10, veran*n*e 4, p*er* il nu*mer*o di quello che **radoppiò** et avanzerà 5 p*er* il nu*mer*o di quello che m*oltipli*chò
rotto "fraction, fractional number" *Ad discipulos* 1.2: Pigli che numero se sia mai fallirà, o sano o **rotto**	**rotto** "fraction, fractional number" *Demostratio i, parte i* [5^{r}] troverai modo et regola a trovare numero intero con **rotto**
sano "rational integer" *Summa* [51^{r}] parti per il ditto numero sano del qual togliesti ditti rotti	**sano** "rational integer" *Demostratio vii, parte i* [13^{v}] Diremo uno altro modo a trovare uno numero generaliter in tucti i modi, o rotto o **sano** non porta ad libitum

TABLE 4 Comparison of Filicaia's mathematical language with Pacioli's (*cont.*)

Luca Pacioli	Piero da Filicaia
schisare "to reduce a fraction by dividing denominator and numerator by the same number" *Summa* [48ᵛ 31–37] Perche ale volte ci piaci con piu basse denominationi dovere dire e assignare ditti rotti over fractioni acio piu facilmente lo intellecto nostro aprenda qual parte over parti sonno de uno intero: per questo ene stato trovato un modo dali antichi pratici quale si chiama **schisare** che in francese si chiama Abrevier [...] Unde dico che **schisare** non vol dire altro se non equalmente partire li doi numeri che fanno el rotto cioe quello di sopra: e quello di sotto della riga	**schisare** "to reduce a fraction by dividing denominator and numerator by the same number" *Demostratio* V, *parte* I [11ᵛ] Dunque caverai di questo 1225 350, resta 875 el quale numero partito per 100 ne viene 8 75/100 che **schisati** sono ¾ apuncto
tenere alle mani "to keep in memory" *De viribus, effecto* XX et tu tacite, per ogni novenario **tieni alle mani** 4, commo pira volte sia detto, alla cui summa giognerai le unità che per li rotti servasti, se l'arai haute;	**tenere alle mani** "to keep in memory" *Demostratio* IV, *parte* I [10ᵛ] Ma se lui penserà 7 ch*e* m*oltiplic*ato p*er* 3 fa 21, p*ar*ti p*er* 2 ne viene 10 ½ che hai rotto. Fallo intero, sirà 11 e p*er* q*ue*sto p*ri*mo rotto **tieni alle mani** uno, m*oltiplic*ato 11 p*er* 3 di novo fa 33, partilo p*er* 2 ne viene 16 ½, fallo intero dirà 17 e p*er* questo secondo rotto **tieni alle mani** 2 e havevi 1 p*er* il p*rimo* rotto, fa 3

2.3 *Terminology of Piero da Filicaia in Comparison with that of Leon Battista Alberti*

Again using a table, we now turn to a comparison of Piero da Filicaia's terminology with that of Leon Battista Alberti.

TABLE 5 Comparison of the terminology of Piero da Filicaia (*Giuochi mathematici*) and Leon Battista Alberti (*Ludi matematici*)

Leon Battista Alberti	Piero da Filicaia
–	**aguto** "nail" [174ʳ] qua*n*do no*n* haveste hasta similm*en*te potreste torre u*no* filo et quello co*n* **due aguti** co*n*ficchare al muro di qua *et* di là.

TABLE 5 Comparison of the terminology of Piero da Filicaia and Leon Battista Alberti (*cont.*)

Leon Battista Alberti	Piero da Filicaia
altezza "height" [4] Se vorrete misurare l'**altezza** d'una torre dove non vi potete appressarvi	**altezza** "height" [168v] Tertia investigatio di uno altro modo di misurar*e* similit*er* una **altezza** di una tor*r*e co*n* l'occhio
–	**asta ferrata** "perch, thin and long wood, with iron tips" [165r] ficchate in terra quivi uno da*r*do dina*n*zi a voi ch*e* stia imobile et fermo overo potete torre una **asta ferrata** a simile artificio deputata, la quale bene si possa in terra fe*r*ma*re*.
battere "to arrive, to mark the point" [1] lì dove il vedere vostro **batte** nel dardo	**battere** "to arrive, to mark the point" [165r] che **batta** el vostro vedere per il dardo
–	**canto** "corner (of the square)" [165r] fermatevi i*n* su dicto **canto** et ficchate in terra quivi uno da*r*do dina*n*zi a voi ch*e* stia imobile et fermo
cera "wax, by extension sign made with the use of wax on arrow or rod" [1] fatevi porre un poco di **cera** per segno, e chiamisi questa cera A	**cera** "wax, by extension sign made with the use of wax on arrow or rod" [165r] fatevi porre uno pocho di cera et chiameremo questa **cera** A
cupo "deep" [8] una profondità d'una acqua molto **cupa**	**cupo** "deep" [174v] *et* el pozzo sino all'aqua sia br*accia* 18 **cupo**
dardo "arrow" [1] poi scostatevi da questo **dardo** quanto pare a voi	**dardo** "arrow" [165r] Et poi schostatevi da questa hasta o **dardo** qua*n*to pare, a duoi o tre o quattro braccia, no*n* importa
diritto del dardo "straight part of the arrow" [1] dirizzando il vostro vedere a mira per el **diritto del dardo**	**diricto dell'asta** "straight part of the pole" [165r] misurare p*er* **el diricto della vostra hasta**

TABLE 5 Comparison of the terminology of Piero da Filicaia and Leon Battista Alberti (*cont.*)

Leon Battista Alberti	Piero da Filicaia
filo piombinato "wire with a piece of lead hanging to determine the vertical of a point" [1] Questo medesimo potete fare con uno **filo a-piombinato**, facendolo pendere dinanzi da voi e segnando le mire vostre con tre perle come altre volte vi mostrai.	**filo piombinato** "wire with a piece of lead hanging to determine the vertical of a point" [166^r] quando non havessi hasta o dardo, potete fare con uno **filo pio*m*binato** face*n*dolo pende*r* innanzi ad voi et segnare in epso le mire vostre co*n* 3 paternostri o perle co*m*me ad voi parrà più a proposito.
cannuccia "perch (made of cane)" [4] tante volte *AB*, cioè tutta **la canna**, misura tutto il profondo del pozzo	**hasta** "perch, thin and long wood" [174^v] tante volte tucta l'**hasta** ente*r*rà nello spatio tucto del pozzo.
scodella "vase, deep dish for soups" [4] o più presto qualche **scodella** piena d'acqua, e ponla in terra, e discostatevi da essa	**infreschatoio** "shallow vase where glasses were put to refresh, refresher" [168^v] o quando no*n* havessi specchio vi servirà uno **infreschatoio** pieno di aqua chiara che omni uno farà el medesimo effecto.
–	**paternostro** "bead (rosary)" [167^r] pigliate uno filo pio*m*binato et fatelo pende*re* dava*n*ti a voi *et* co*n* perle o **pate*r*nostri** signate li puncti
profondità "profundity" [8] Con questi perfino a qui recitati modi di misurare, potete simile misurare ogni **profondità**, ma per essemplo ne porremo qualche modo certo.	**profundità** "profundity" [173^r] Septima investigatio a misurare una **profundità** d'uno pozzo o simili
ragione "ratio, proportion" [2] troverete pari **ragione** fra *CE* e *DE*	**ragione** "ratio, proportion" [166^r] Troverrete nel vost*r*o dardo simile **ragione**, cioè che com*m*e quella pa*r*te ad voi nota della tor*r*e, entra
cima "top" [1] Chiamisi **la cima** della torre *A*	**sommità, cima** "top" [166^v] quanto pare ad voi co*n* modo poter*e* bene veder*e* la sua so*m*mità [167^r] *et* chiamerete **la cima** del dardo A

TABLE 5 Comparison of the terminology of Piero da Filicaia and Leon Battista Alberti (*cont.*)

Leon Battista Alberti	Piero da Filicaia
specchio "mirror"	**specchio** "mirror"
[4] Chiamisi la cima della torre *A* e il suo piè B, lo **specchio** C, l'occhio D, e il sito vostro dove sono e' vostri piedi si chiami E,	[169r] Sia la cima della tor*r*e A, el piè suo B *et* lo **specchio** sia C *et* il v*ost*ro occhio sia D *et* li vostri piedi sieno E.

CHAPTER 4

The Cultural and Literary Relevance of the Manuscript

1 Cultural References

The manuscript analysis was also significant in terms of cultural aspects. In addition to the dedication to Giuliano, we find other interesting evidence in it, such as a quotation from Pulci, some geographical references to Florence and its surroundings, and even some objects used for playful purposes at court, such as *crazie* or the typical Florentine cards called *germini* or *trionfi* (Palmarini 2021, 214–222). Furthermore, as previously stated, during the analysis of the Filicaia manuscript, a number of anecdotes were discovered that are of some interest due to the manner in which they are presented. One of them in particular stands out.

2 Columbus's Egg

Almost everyone has heard the story of Columbus's egg, which has become proverbial. A closer examination reveals that the anecdote has several literary attestations from the 16th century, including Benzoni (who attributes the story to Columbus), Vasari (who attributes it to Brunelleschi), and Pacioli in *De viribus quantitatis* (attributed to Brunelleschi). The treatise, written shortly after Pacioli's work, contains a different, diverse, and richly detailed version of the anecdote. Filicaia proposes this story as an ancestral tale and describes the entire incident in detail. It is also important to note that the story is not an end in itself, but serves a specific purpose in the treatise: it assists the author in claiming the originality of the collection of mathematical riddles and the creativity of the Florentine genius.

Here is the story of the egg as told in our treatise (ff. 44^{v}–47^{r}):

> Ricordami havere udito dire, Magnifico Juliano, dalli mia antecessori et antiqui che intervenne nelli preteriti tempi allo Illustrissimo duca di Milano volere fabricare un ponte in Lombardia sopra una fiumana decta l'Adda, se bene me ricorda, et a ciò meglio tale effecto conseguire potessi, convocati tucti li maestri et architectori della sua Lonbarda regione, pro-

 | DOI:10.1163/9789004727984_005

posta loro tale opera volere mandare ad effecto che desiderava, li dicessino l'ordine et modo che a tale artificio era necessario et che con mancho spendio et più facilità che alloro fussi possibile li dessino disegno. Il che inteso dalli prenominati maestri et architectori et sopra ciò preso quella conclusione li parve oportuna, mostrorono alla excelentia del signor Duca tal cosa non potersi fare senza difficultà et per questo ne seguitava spese a tale opera inconvenienti. Il che inteso el Magnifico duca di tale cosa sbigottito più et più anni la lasciò inperfecta comme quello che giudicava essere quasi impossibile. Advenne che, forte fortuna, capitò nella ciptà di Milano uno architectore fiorentino il cui nome adesso non l'ho alla memoria et, essendo epso nella prefata ciptà, el duca comme desideroso de intendere et vedere li homini virtuosi un giorno lo fece ad sé venire et, fra li altri electi ragionamenti, li dimandò consulta sopra questo ponte che più anni comme iusto et bono signore et di sui popoli amorevoli haveva desiderato di fabricare perché a popoli sua viandanti dava grandissima incomodità questo fiume. El nostro architectore fiorentino, inteso et visto con lo spiculativo occhio la cosa di che qualità era, brevemente li respose che con mancho spendio et con molta più facilità tale opera fabricare si poteva che dalli sui architectori li era stato exposto offerendosi lui in simile exercitio volentieri affatigarsi per fare cosa grata a sua signoria. El che molto piacque al duca et, aceptate le sue offerte, dette principio a fabricare uno magno et spatioso ponte sopra el predicto fiume. Hora, comme adviene, el nome andò per tucta la Lonbardia comme uno maestro fiorentino haveva principiato el prenominato ponte. Onde, molti et vari architectori dilectandosi intendere vennono a vedere el modo e l'ordine che costui haveva trovato. Di che naque che molti o la magior parte inteso el priegio che costui aveva di dicta opera, andorno al signor duca dicendo che con mancho assai harebbono loro tale ponte fabbricato che non faceva questo fiorentino, il che indusse nell'animo del duca un pocho di sdegno che si riputava giunctato et alquanto inanimito contra epso maestro. Et, factolo a se venire, li disse havere molti maestri et architectori nella sua ciptà che questo ponte harieno per mancho priegio assai che lui conposto et fabricato di che lui intendeva che lui venisse a pregi che e sua architectori erano consueti. Donde el nostro architectore, comme homo experto di acutissimo ingegno, dimandò in gratia al duca che facessi venire quelli homini che di simile cosa si vantavano che voleva con loro conferirse et non era per iscostarsi dalle cose iuste. Et ragionevole il che subitamente fu facto. Venuti adunque li prenominati maestri et constituiti nella presentia dello Illustrissimo signore duca insiemi col nostro fiorentino, furono i predicti da epso dimandati se erano loro quelli che tale cosa admiranda fare volevano il che dal loro fu quivi confirmato. Onde el nostro fiorentino, preso in mano uno uovo et postolo insù uno schacchiere pulito, disse alta voce: horsù, chi di voi ferma questo uovo qui nel mezo di questo schacchiere senza romperlo et con presteza cioè innanzi che passi uno quarto di hora hoggi, guadagnerà 50 ducati. Onde tucti stupiti quivi non fu nessuno che potessi questo effecto mandare ad executione, ma guardando l'uno l'altro si ristrignevono nelle spalle. Onde, passato lo spatio da llui dato, che passò presto, disse loro: se la rizassi io nel modo predicto, voi direte bene che io sia valente homo? Tucti una voce dissono: si certo perché in facto chi non ha el modo et sua facilità li parrà cosa stupenda, onde furno constrecti dire sì. Et lui, preso di uno cantuccio o camino uno pocho di cenere et quella messa in sul mezo dello schacchiere, vi rizzò su l'uovo con la sua puncta, onde li maestri lombardi cominciarno a ridere facendosi beffe di simile acto. Dissono che in quale modo anchora loro saprebbono porre l'uovo, di che el nostro fiorentino rispose: così sapete fare el ponte hora ch'io vi ho mostro el modo, ma prima ch'io vi mostrassi la via non sapevi fabbricare el ponte et non sapevi rizare l'uovo nel mezo dello schacchiere et omni uno sa fare a giuocho insegnato. Alle quali parolle non fecero risposta alcuna. Il che inteso el magnanimo Duca molto li piacque simile acto et giudichollo homo veramente di mirabilissimo ingenio et, licentiati li sua architectori, li disse che alla sua opera attendessi. Ho voluto, Magnifico Iuliano, questa brieve faccetia recitare perché forse sarà qualche uno più presunptuoso ch'el bisogno che dirà: io anchora havrei saputo una simile opera compilare. Et diranno questa essere suta di grandissima facilità a che rispondo comme el nostro fiorentino che ciaschuno sa fare a giocho insegnato et che quando la via è facta omni uno sa per epsa facilmente caminare.

I recall hearing from my ancestors and other elders that some time ago, Magnificent Giuliano, the illustrious duke of Milan wanted to build a bridge in Lombardy over a river called Adda, if I recall correctly, and in order to better achieve this goal, he summoned all the masters and architects of his Lombardy, proposing that they carry out this work and asking them to tell him the order and the way that was necessary for this work and when the aforementioned masters and architects heard this, they devised a solution that seemed appropriate to them and demonstrated to the Excellency Lord Duke that such a task could not be completed without difficulty and thus would result in excessive expense. When the magnificent duke realized this, he was shocked and abandoned the project because he thought it was nearly impossible to complete. It so happened that a Florentine architect, whose name I no longer recall, happened to be in the city of Milan, and while he was there, he was summoned by the duke, who liked to meet and talk with virtuous men. Among other things, the duke asked him for advice on the bridge, which he had wanted to build for years as a just and good prince and lover of his people because the river was bothering people so much.

Our Florentine architect, having understood and seen with his speculative eye what it was all about, replied briefly that this work could be realized much more easily and affordably than his architects had presented to him. And he offered to do such a task in order to show his gratitude to the prince. This pleased the duke greatly, and after accepting his offer, he began construction of a large and spacious bridge across the river. As is often the case, word spread throughout Lombardy that a Florentine master was working on the bridge. As a result, many different architects who wanted to understand came to examine the manner and order he had discovered. As a result, when many or most of them learned of the fee he was to be paid for that work, they went to the lord duke and told him that they would build the bridge for less money than this Florentine. This infuriated the duke, who felt cheated and began to hold a grudge against the aforementioned Florentine master. So he summoned him and informed him that he had many masters and architects in his city who would build this bridge at a much lower price than his, and by doing so, the duke intended for the master to reduce the price to the level of the duke's architects.

At this point, our architect, an experienced man of keen intelligence, requested a favour from the duke. He asked him to bring in all the men who boasted of such a thing because he wanted to talk to them about it, not to avoid doing what was right and rational. This was done right away. So, when the aforementioned masters arrived, he asked them, in the presence of the most illustrious lord duke gathered with our Florentine, if it was they who wished to do an admirable thing, and they confirmed this. Then, taking an egg in his hand and placing it on a clean chessboard, our Florentine declared aloud: "Whoever stops this egg straight upright here in the middle of this chessboard without breaking it and quickly, that is, within a quarter of an hour, will earn 50 ducats."

Everyone was amazed and no one could complete this task, but when they looked at each other, they shrank back. Then, when the time he specified had passed (and it did), he asked them, "Would you say I was good if I put it in the manner predicted?" They all said, "Of course," because to those who do not have the way or the facility, it will appear to be a

> wonderful thing, so they were forced to say yes. And he took some ash from a fireplace or hearth and placed it in the centre of the chessboard, and he placed the egg with its point upwards, and the Lombard masters laughed, mocking such an act. They claimed that they were also able to place the egg this way, to which our Florentine responded, "So you know how to make the bridge now that I've shown you the way, but before I showed you, you didn't know how to build the bridge and you didn't know how to place the egg in the middle of the chessboard." When the game is taught, everyone knows how to play it. They were unable to respond to those words.
>
> The magnanimous duke, who had witnessed the entire thing, was very pleased with our architect's act and thought him to be a man of great brilliance, so when his Lombard architects left, he told the Florentine architect to get on with the construction. I wanted to tell you this brief anecdote because, perhaps, someone more arrogant than necessary will say, "I, too, would have known how to complete such a work." And they will claim that it was extremely simple. Like our Florentine architect, I respond that everyone knows how to play the game once taught and that once the path is built, everyone knows how to walk on it easily.

In Filicaia's version of the anecdote, the querelle, which will result in a demonstration of how to make an egg stand upright, thus presents significant differences from the versions by Pacioli (earlier), Vasari, and Benzoni (later). The story is set at the court of Milan and concerns the difficulties encountered in the construction of a bridge across the Adda River by builders in the service of the duke. The protagonist of the tale is a Florentine architect who, in order to assert the creativity of Florentine genius among his Lombard colleagues, demonstrates to those present how he can make an egg stand upright. The act takes place on a chessboard and, moreover, the eggshell is not crushed as in other tales, but the trick is to rest it on a mound of ash. In addition, we have identified the character of the architect who is the protagonist of the tale in Filicaia's version. Although the author never names him explicitly, various clues, including chronology and a probable location of which bridge it might be, lead to a 15th-century Florentine architect, Antonio Averlino, known as Filarete (Palmarini and Sosnowski 2020, 175–178) who lived in Milan in the period between 1451 and 1466. Thus, Columbus's egg turns into Filarete's egg. Filicaia seizes the opportunity to extol the Florentines' creativity and genius once more.

Conclusions

Piero di Filicaia's manuscript *Giuochi mathematici* reveals the multifaceted significance of Renaissance intellectual culture, particularly within the fields of philology and history of mathematics, through an in-depth analysis. We gain a deeper comprehension of the mathematical and recreational activities that took place at Renaissance courts, and the manuscript also sheds light on the linguistic and cultural dynamics that were prevalent at that time. Piero di Filicaia's work makes a significant contribution to the history of mathematics. It exemplifies the lively mathematical recreation that took place in Florence, in the de' Medici *milieu*, during the Renaissance period. Filiacaia's manuscript, which is full of games, puzzles, and riddles, exemplifies the popularity of such entertainments among the elite as well as their intellectual prestige. These pursuits were not merely a form of leisure; rather, they were an essential component of the intellectual exercises that were prevalent in this period, serving both educational and social purposes. Mathematical games were able to facilitate the dissemination of numerical skills and logical thinking among the social elite, which in turn fostered a culture of learned amusement that was both sophisticated and accessible. In order to demonstrate his originality and intellectual agility, Filicaia's creative approach to adapting and expanding upon previously established mathematical problems is quite impressive. The fact that he is able to incorporate intricate mathematical ideas into the fabric of everyday entertainment demonstrates that he has a profound understanding of the preferences of his audience as well as the educational objectives of the time period. This adaptability not only maintains the manuscript's relevance as a historical document, but also positions it as a valuable source of rich material for contemporary historians of mathematics, comparable to Luca Pacioli's more famous work *De viribus quantitatis*.

Filicaia's manuscript is a treasure trove from a linguistic point of view. His deliberate use of a refined Florentine dialect indicates a nuanced assertion of local identity and intellectual independence. Despite his interactions with Luca Pacioli, the Paciolian language does not appear to have influenced him, despite their shared contents and ideas. This linguistic choice serves as a cultural statement, bringing Filicaia into alignment with Florence's intellectual tradition and establishing the city's cultural autonomy within the larger Italian linguistic landscape.

In addition to this, the manuscript offers a wealth of information regarding the transmission and transformation of specialized vocabulary among Renaissance scholars. In the process of drawing inspiration from both Pacioli and

 | DOI:10.1163/9789004727984_006

Alberti, Filicaia takes part in a lively exchange of ideas that contributes to a deeper comprehension of the intellectual currents that were circulating in Florence during that historic period. This intertextual dialogue highlights the fluidity of knowledge exchange as well as the role of individual agency in the development of conceptual and linguistic frameworks.

The examination of the manuscript not only enhances our comprehension of the social and cultural mechanisms that facilitated the dissemination of mathematical knowledge, but also deepens our appreciation of Renaissance courtly life, which was characterized by a complex combination of leisure, education, and intellectual competition. The manuscript describes convivial settings that show how mathematics permeated social interactions and cultural practices. This serves to reinforce the interconnectedness of knowledge, power, and social life in Renaissance Italy.

In conclusion, Piero di Filicaia's *Giuochi mathematici* is an important component in our comprehension of the intellectual and cultural *milieu* that existed during the Renaissance. It is not only a reflection of the era's characteristic combination of entertainment and education, but also contributes to the ongoing academic discourse on the development of mathematical thought, linguistic expression, and cultural identity in early modern Europe. In light of this, it encourages additional academic research that cuts across a variety of fields, which holds the promise of revealing fresh insights into the intricate web of intellectual life during the Renaissance.

Editorial Policy and Principles

We have based the edition on the text of K, trying to ensure the text remains as accurate as possible. We accept readings from F where the readings in K are clearly faulty. Every time the readings in K and F are different, we indicate in the apparatus the reading that we did not accept.

However, we do not consider different graphic uses or tiny morphological oscillations as different readings. Because we consider K an "official" copy, closer to authorial intent, in the rare cases where there are differences in linguistic forms between manuscripts, we follow K. It is worth noting that these small differences are transversal, which means that different morphological forms exist both within and between manuscripts. However, there are no forms unique to a manuscript, and the distribution of oscillations does not appear to follow any particular pattern. To give an example, *fussi* and *fusse* for the third-person singular of being are both present, with *fussi* taking precedence. In the first section of the treatise, there are nine occurrences of *fussi* and two of *fusse*. The manuscripts show no significant preference for one form over the other. Other oscillations exhibit similar dynamics, so we omit their reporting in the apparatus in order to focus solely on philological variants.

We adhered to current philological practices when making editorial interventions with a conservative approach. When scribes use "j" for the vowel /i/, we use "i", but for the semi-vowel, yod, we use "j". The letter "u" represents the vowel /u/ while "v" represents the labio-dental voiced fricative /v/. Punctuation was added to clarify syntactical relationships, but it has not been standardized to match modern implementations.

In the edition, we dissolve standard abbreviations, relying on what the scribes used in the full forms of the abbreviated words. Despite our conviction that the edition should adhere to very conservative criteria, we encountered difficulties in transcribing numbers, particularly those referring to the numbering of individual units of the text (*demostrationes, casi, investigationes*). In this case, faced with a variety of solutions adopted by the scribes (numerals written in letters, Roman numerals, Arabic numerals, and various intermediate solutions), we decided to keep the solutions as close as possible to the original.

Two other number conventions that we felt obliged to adopt should also be noted. In order to ensure that the edition is as functional as possible for those looking for mathematical information in the text, we transcribe the ordinal numerals in the text of the riddles in letters (*primo, secondo, tertio, quarto, quinto, sexto, septimo, octavo, nono, decimo*).

 | DOI:10.1163/9789004727984_007

Another issue was the transcription of fractions. We adopted the slash in order to standardize mathematical reading (1/2 for "a half", 1/4 for "a quarter", etc.).

Giuochi mathematici

DOI:10.1163/9789004727984_008

Piero di Niccolò d'Antonio da Filicaia

Giuochi mathematici

Giuochi mathematici: Repertorio et tabula Piero di Niccolò d'Antonio da Filicaia 11r

Repertorio et tabula della seguente opera
et libro dicto Giuochi mathematici facto et conposto
Per Piero di Nicolao d'Antonio
Da Filichaia da Firenze, dedicato
allo Illustrissimo Magnifico Juliano
de' Medici, suo
observandissimo

vi^{r} *Fratris Paschasii Burgensis ordinis minorum*
Petri auctoris infrascripti in pueritia praeceptoris
ad librum epigramma

I, liber, i foelix Medicum visure penates,
i pro me et domino dic reverenter: "Ave"!
Post hec subiunges illi mea paucula verba:
Petrus te hoc donat, stirps Philicara, libro.

Eiusdem fratris Paschasii Burg[*ensis*] *epigr*[*amma*] *Ad Magnificentissimum Julianum Medicen*

Hunc librum generose Juliane,
cun[*c*]*tis Magnificentior, parentis*
Laurenti soboles, decusque avorum
Sublime et Medicum reversionis
Auctor, praesidium bonis tuisque
Donamus tibi sicque destinamus,
Qui cum de numeris loqui iocosis
Seu ludis numeralibus valebis
Si quando vacuum manere detur.

O Rector patrie, presidium decusque VI^v
O secli columen fluentieque
Quo, te sospite letiore vita
Stat Florentia civitas beata
Foelix nec metuet Leo ruinas
Bellorum Italiae tumultuantis.
Vive ergo atque vale, librumque nostrum
Humane accipias beatiorem
Quem scimus fieri manu potentis.

Dedica

1r *Ad Magnificum ac Clarissimum virum Julianum Medicen Patritium generosissimum Florentinum ac Dominum suum colendissimum Petri Nicolai Antonii de Filicharia prefatio incipit foeliciter.*

Considerando in me medesimo, Magnifico Juliano Medice, di quanta utilità, di quanto proficto et dignità sia la scientia mathematicha, quella sommamente laudare, honorare et sopra tucte le altre scientie exaltare dobbiamo, con ciò sia cosa che niuna altra scientia habbia in sé sì vera et unica perfectione.

Perché, comme dice el philosopho, omnia numero, pondere et mensura continentur: vole significare che tucte le cose create in numero, peso et misura consistono. Et questo veggiamo essere la verità che nulla, nel mondo o nel cielo, si trova che in sé non habbi parte delle prenominate cose. Possiamo anchor dire questa digna scientia havere in sé qualche parte di divinità che 'l sia la verità
1v comme | la sancta et individua trinità è infinita, così questa scientia procede in numero et in misura infinita.

Et, comme la sanctissima et individua trinità è infallibile, cioè che mai non può fallire, così la scientia mathematicha è infallibile ché, data in questa scientia una vera regola et quella bene operando, non può fallire. Per questa tucta preallegata ragione possiamo et dobbiamo dire veramente questa scientia esser divina et in omni genere perfecta. Potrei molte altre ragioni et auctorità in laude di tale scientia allegare, ma questo basti. Atteso io adunque tucte le sopradicte cose nelli mia giovenili anni, molto mi sono dilectato a tale scientia dare opera et ho visto, lecto et rivoltato molti auctori sopra tale materia parlanti. Fra quali el nostro perspicacissimo Euchlide Megarese molto difusamente ha parlato et certo molto a tale auctore siamo obligati inperoché questa dignissima scientia era quasi annullata et ad nihilum redacta et per le virtù et eficacissime propositioni et documenti di tale homo hoggi è pervenuta in grandissima con-
2r ditione et sopra epso Euchlide mol|ti homini di laude digni hanno scripte et conposte molte varie expositioni, comme son certo Vostra Magnificentia Illustrissima apertamente sa et cognosce. Io adunque, havendo molto dato opera a tale facultà, ho inteso molti secreti, forze et virtù di quella, quali tucti ho insiemi riducti et adunati et ho conposto et fabbricato questo libretto, opera in verità digna et dilectabile, el quale si chiama «De ludis mathematicis», cioè «Di giuochi mathematici». Vederà, Vostra Magnificentia prefata, in questa nostra opera molti secreti di decta scientia con e' quali per virtù di numero et misura

---

14 K] divintà 17 K] et et 25 K] perspicassimo 32 F] ei quali

poterà pervenire alla notitia di varie et diverse cose le quali non passeranno senza grandissima dilectatione et piacere di animo et mente di quella come legendo poterà vedere et apertamente conphrendere.

Et bene mi è noto la bona et fida servitù della felice anima di messer Bartholomeo Scala in verso el vostro dignissimo padre—ai quali Idio per sua pietà doni pace et requie sempiterna—dal quale messer Bartholomeo per materno genere sono disceso, el quale ciaschuno sa comme fu tanto exaltato dalla felicissima anima del Magnifico | Lorenzo per le sue virtù che possiamo et 2v
dobbiamo dire el prefato Messer Bartolomeo essere homo per li meriti del Magnifico Lorenzo in verso di lui, et benché el prefato Messer Bartolomeo per le innumerabili sue virtù meritassi essere exaltato, non fu mancho laudabile l'opera del Magnifico Lorenzo in questa parte a vedere et cognoscere dicta virtù perché come dice el philosopho:

Illos quidem summa laude dignos arbitramur qui virtutem quocumque loco sit facile perscrutantur virtus nam incognita frustatorie appellari solemus. Dice Aristotele che colui è digno di laude che cognosce la virtù facilmente in qualunche loco la si sia perché la virtù incognita è quasi niente che sia il vero. Che varrebbono le virtù medicinali delle herbe et altre cose se li antiqui doctori di medicina non le havessino cognosciute et datone notitia? Che varrebbono le virtù delle pietre preciose se li predecessori nostri non havessino tali virtù cognosciute et dato lume? Et così si può dire di tucte le virtù che dalli famosi homini state intese et cognosciute sono.

Concludiamo adunque non essere laudabile mancho el Magnifico Lorenzo in questa parte in havere tanto perfectamente cognosciuta la v|irtù del prefato 3r
Messer Bartholomeo che lui proprio. Et desiderando io immitare in qualche minima parte epso Messer Bartholomeo, deliberai questa mia opera a Vostra Illustrissima Magnificentia dicare in memoria et firmo testimonio della mia vera et fida servitù in verso di quella perché cognosca epsa amatore di tucte le virtù et di simili cose molto dilectarsi. Supplico adunque a Vostra Magnificentia prefata si degni tale opera benignamente acceptare et me nel numero delli humili et fidi servitori di quella collocare, alla quale del continuo racomandomi *quam Deus omnipotens in summa felicitate ac gloria acrescere manutenere ac conservare dignetur.*

---

16 K] vrtù

Prologus incipit

Molti varii auctori, digni di laude, nelli preteriti tempi hanno scripto et con-
posto diversi et varii volumi sopra la mathematicha disciplina et quali si sono
forzati con omni industria et ingenio darci più vera notitia et cognitione di
tucti i sua primi fundamenti et operatione che suto possibile. Fra quali el padre
delle scientie mathematiche Euchlide Megharese, philosopho dignissimo Boe-
3v tio, Archimede et mo|lti altri homini in tale scientia peritissimi hanno parlato
molto largamente sopra la prefata scientia et di sua proportione et proportio-
nalità hanno toccho cose alte et profunde. Oltra questo poi hanno scripto molti
altri nella materia più bassa, cioè nel fare ragione di merchantie et ragione di
argento et oro et misure et pesi comme a ciaschuno è noto. Hanno tanto sopra
tal materia parlato che a me parebbe superfluo et vano volermi in simile opera
afatigare, ché hora mai el partire et multiplicare, racorre et sottrare ordinaria-
mente et rotti et interi et l'ordine et modo di fare ragioni ordinarie, comme
meritare, consolare et ricare a uno dì, è tanto noto et manifesto a tucti et è sì
in tale modo promulgato che a me pare conveniente el preterirla. Ma vogliamo
in questa nostra opera tractare et dimostrare cose sottile et, sino a questo dì,
note a pochi, cioè secreti della virtù et forza della quantità continua comme
discreta ché per forza di multiplichatione et subtractione et partimenti possia-
4r mo venire alla notitia di molte bellissime | cose, comme per exemplo Vostra
Magnificentia Illustrissima poterà conprehendere.

Potrebbe qui dire alcuno: "Piero, sta' fermo uno pocho. Onde hai tu tracto questi tui secreti? Tu li hai rubbati?" A che rispondo brevemente che niuno in questo mondo è indovino né equale allo omnipotente Idio che omni cosa sa et cognosce.

Et pertanto, se io ho racolto di più luoghi questi mia secreti et forze et forse parte rubbati et a questo et a quello, non sia in questa parte nessuno che me riphrenda, perché comme dice el tuo poeta fiorentino Luigi Pulci nel fine di una stanza che chi non rubba è chiamato ribaldo.

Et bene, è virtù et ingenio mirabile et a suo proposito el sapere rubbare et
acomodare nelle cose virtuose et, che sia il vero, hoggi dì si dice: a rubbare uno
buono cane o uno buono ucello non è peccato. Et aciò di me nessuno si possa
iustamente dolere, lo dico aperto a tucti che da me si guardi ciaschuno ché,
quando in simili cose virtuose potrò alcuna cosa rubbare, non ci harò adver-
4v tentia alcuna. Ha! io non ho perhò rubbato! Vedi che già lo crederesti et io |
me ne farei grande conscientia. Conclusive parlando: *nihil dicitur quod non sit*

---

30 K] el sapere a suo proposito rubbare

dictum prius. Dirò adunque in questa nostra opera alcune virtù di dicta quantità, cose invero dilectevole et bellissime et da havere di epse mirabilissima consolatione et piacere perché farai molti giochi stupendi apresso a ciaschuno che non havessi vera notitia di questa opera perché sino a hoggi non è chi di tali secreti habbi scripto et composto pubblicamente. Distinguerò adunque questi nostri capituli dicendo: Prima demostratio, Secunda demostratio, Tertia, Quarta, etc. Et questa opera meritamente si può chiamare uno gioiello della scientia mathematicha. Vegniamo adunque alla conclusione. Dirò così: et prima etc.

Prima parte del libro di giuochi mathematici

Prima demostratio a trovare uno numero intero

Prima et principalmente voglio dare modo a trovare uno numero intero che per uno fussi stato pensato nella fantasia. Ovvero uno dicessi: "io ho in mano una quantità di danari o di quarteruoli o chiose" o quello vuoi che non importa,
5r purché non | vi sia rotto, cioè non sia mezo o tertio, ché quando vi fusse rotto daremmo di sotto altri modi. Questa prima regola serve a trovare uno numero intero comme è dicto qui a piè. Seguendo di leggiere troverai modo et regola a trovare numero intero con rotto et rotto solo, et tucti e' modi. Le quale cose tucte si fanno per virtù di numero. Diremo adunque così: uno ha preso uno numero intero nel quore suo. A volerlo investigare farai così: prima di' che sopra tucto il numero vi pongha su la metà di quanto prese et dimandali se in questa metà vi viene rotto. Cioè se lui havessi preso 7 volendovi mettere su la metà, sarebbe tre et mezo, dicendo lui havere rotto, lo facci intero, cioè se ha tre et mezo, dica 4. Et questo giungha sopra 7, fa 11. Et tu per questo rotto la prima volta terrai a mente da te uno. Quando non viene rotto, cioè che lo amico havessi preso pari, non vi è disputa et non hai a tenere a mente nulla. Dipoi dirai una altra volta che prenda la metà di quanto si trova in tucto et ponghalo in su la somma. Si trova anchora similmente dimanda se in questa seconda metà
5v vi viene rotto. Se non | viene rotto, non importa, ma dicendoti esservi rotto, dirai lo facci intero. Et per questo secondo rotto tieni alle mani 2; et del primo rotto havevi uno, fanno 3. Che questo mectendo sopra 11, la sua metà è 5 et 1/2, che facendolo intero sono 6, che giugnendolo sopra 11 sono 17. Et questo facto, dimanda: in tucta la somma, quante volte vi entra 9? Et per omni volta entra 9, tu tacitamente tieni alle mani 4, cioè se v'entra una volta, tieni 4, et se dua volte, tieni 8, se tre, tieni 12 et così in infinito. Et sopra quello poni quello havevi alle mani. Mai falla. Comme in questo hai per tucta la somma 17 che comme si vede partita per 9, ne viene 1, avanza 8. Ma quello che avanza non importa. Dirò adunque che in 17 non entra più che una volta per la quale volta ho a tenere a mente 4. Et prima havevo alle mani 3 che, posto sopra 4, fa 7. Che tanto dirai che lo amico si ponessi in numero, cioè 7. Et nota che la prima volta se viene rotto hai a tenere a mente uno et, s'el rotto non è alla seconda, non porta. Per la prima sempre tieni uno. Et s'el rotto solamente fussi alla seconda et non alla prima,
6r sempre alla seconda volta havendo | rotto, tieni a mente 2. Et se non è rotto né al'una né all'altra, non hai a tenere a mente nulla, se non per omni 9 che entra

---

1 prima parte del libro di giuochi mathematici] F *omissis* 5 quarteruoli] F quartiroli 14 rotto] F rotto dirai 30 dirai che lo amico] K dirai lo amico

nella somma dello amicho. Et tu di tucti tieni 4. Comme se lo amico pensassi 8, dirai vi ponga su la metà che è 4, fa 12, che non ci è rotto. Anchora dirai sopra 12 vi pongha la sua metà che è 6, che anchora non ci è rotto, farà 18. Domandali quante volte vi entra 9. Dirà: "due volte". Adunque dirai che in sua mente lui prese octo. Et se lui avesse preso 6, postovi la metà, fa 9. Et la prima volta non ha rotto, et sopra 9 posta la metà—è 4 1/2—che vi è rotto alla seconda volta, dirai lo facci intero, fa 5. Posto sopra 9 fa 14. Et tu harai alle mani 2. Poi dimanda quante volte in la sua somma entri 9. Dirà "una" et tu dirai 4, et 2 havevo alle mani, fa 6. Et tanto si pose in quore. Et se pigliassi 5, che la meta è 2 1/2, dirai lo facci intero et pongalo sopra 5, farà 5 et 3, fa 8. Et hai la prima volta rotto et la seconda non hai rotto, che la metà di 8 è 4, che, posto sopra 8, fa 12, che v'entra una volta 9, hai 4 et uno alle mani, fa 5 et tanto si pose l'amico. | Et se l'amico 6v
havessi pensato 3, postovi su la sua metà—che è uno et mezo—fallo intero, et fa 2 et fa 5. Et tu per questo primo rotto hai alle mani uno et poi di sopra 5 poni la sua metà—è 2 1/2—fallo intero—è 3—fa 8. Et tu per questo secondo rotto tieni a mente 2 et uno havevi prima, fa 3. Dimandando quante volte vi entra 9, dirà: "niuna" et tu dunque dirai si pose 3 che hai alle mani. Hai lo exemplo et modi. Ti può venire o rotto la prima volta et non la seconda, o rotto la seconda et non la prima o rotto seconda et prima, et non rotto mai. Et non ascendendo el tucto alla somma di 9, potrai per questa prima dimostrazione fare questo bellissimo giuocho. Solo ti dico che, quando lo fai, lo facci più coperto puoi, aciò non se ne avegghino li circumstanti perché si trova alle volte mirabili ingenii che, stando a vedere fare uno giocho, di facto conphrendono el modo. Perhò ti advertischo, Lectore, che li facci copertamente, che quando tu dimandi quante volte entra 9 nella tua somma, lo potrai investigare in varii modi comme per te potrai conphrendere. Hora nota uno altro modo quasi simile al primo.

Secunda demostratio a trovare uno numero intero senza rotto 7r

Potrai per uno altro modo bello venire alla medesima notitia. Et questo modo ti servirà aciò le brighate non possino comphrendere la regola tua che, comme interviene alle volte alle brighate, tu fai uno giuocho che piace a tucti: chi per dilecto et chi per veder di conphrender la regola tua, dicono fallo una altra volta. Fallo in modo ti bisogna compiacere a lloro, aciò niuno possa conphrendere tu muta regola et modo. Per questa seconda regola farai così: cioè quando l'amico harà pensato uno numero senza rotto che essendovi rotto questa regola non serve, pertanto dirai allo amico tuo che quello numero pensato lo multiplichi per 3 et quello viene, dirai ne facci 2 parte magiore che si può. S'el numero

23 comphrendono] F comphrendendo

sarà pari, saranno 2 parti equali, s'el sirà caffo, seranno una magiore che l'altra. Comme se lo amico dirà haver preso 6, multiplicato per 3 fa 18, facendone 2 parte magiore si può, sirà 9 per parte, allora dirai che una altra volta lo multi-
7v plichi per 3 che in questo farà 3 via 9 fa 27. Allora comme di sopra dimanderai | acortamente sottraendo lo amico quante volte entra 9 nella sua ultima somma, il che saputo, per omni 9 terrai alle mani 2, adunque in 27 che lui haverà alle mani vi entra 3 volte. Dirai adunque che l'amico prese 2 via 3 fa 6 et mai ti fallirà questa regola. Ma quando el numero sarà caffo, cioè el numero che piglia la prima volta terrai questo medesimo modo. Cioè diciamo se l'amico pigliassi 7, multiplicha per 3 fa 21, fanne due parte magiore si può, l'una fia 10 l'altra 11. In questo caso quando le parte non sono equali habbi advertentia che sempre bisogna pigli la magiore. Et perhò dirai che la magior parte che lui ha multiplichi per 3, che multiplicherà 11 farà 33, che vi entra 3 volte 9 et avanza 6. Nota che a te bisogna saper quante volte entra 9 nel suo ultimo producto et se avanza nulla, ma non te inporta saper quanto avanza perché sempre avanzerà 6 se el numero sia disparo et tu per 6 che avanzerà hai a tenere alle mani uno, dunque
8r in 33 entra 3 volte 9 che hai a tener a mente 6 et uno per el 6 che avanza fa 7. | Et mai simile regula fallirà. Ma se l'amico pensassi uno, tu dirai lo multiplichi per 3 fa 3, factone 2 parte, la magior parte fia 2, l'altra uno. Presa la magiore, multiplicha per 3 fa 6, dimanda quante volte v'entra 9, dirà "niuna". Dunque dirai prese uno per 6 et questo nasce per questa forza che l'unità sola multiplichata et partita nel modo sopradicto produce 6 comme per exemplo puoi vedere. Et se lo amico havessi preso 2, multiplichato per 3 fa 6, factone 2 parte hai 3 per parte, multiplicato per 3 fa 9 apuncto, parti per 9 vene uno, cioè v'entra una volta. Adunque dirai che pensò 2. Et questa regula fia generale in tucti e' numeri che uno potessi immaginare, pure che la prima volta che lui piglia non vi sia rotto. Et perché io dica fa 2 parte le magiore si può, potrai anchora dire: lo divida per mezo et questo sopradicto modo è dicto per chi non sa travagliare rotti, onde se l'amico piglia 9 multiplicato per 3 fa 27, la cui metà è 13 et 1/2, multiplicato per 3 fa 40 et 1/2, che v'entra 4 volte 9, dirai 8 et uno per 4 et 1/2
8v che avanza fa 9 et tanto prese | et mai falla. Puoi operare quello modo più ti piaci, ma perché omni uno non sa bene travagliare di rotti, userai più presto el modo a dire: "fa 2 parte le magiori si può et piglia la magiore et multiplicha per 3 comme è dicto". Et questo basti quanto alla seconda dimostratione.

Demostratio tertia a trovare uno numero intero senza rotto, quando el numero non sia con rotto, cioè:

Potrai per uno altro modo anchora pervenire alla medesima notitia. Dirai allo amico che 'l suo numero multiplichi per 3 et poi la parta per 2 et se nel partire in 2 ne venisse rotto, dirai lo lassi andare et che quella metà di novo la multiplichi

per 3 et quello ne viene parta per 2 et, se ha rotto nel partire, dirai lo lassi andare et poi ti dica quante volte entra 9 nella sua somma. Et per omni volta che v'entra 9 hai a tenere alle mani 4. Et nota che quando troverai rotto nella prima volta solamente et non nella seconda, allhora dicendo lo lassi andare et tu tieni da te a mente tre. Et quando el rotto non fussi la prima volta et fussi la seconda solamente, allora in qu|ello caso lo lassi andare, dirai et tieni alle mani 2 et quando 9r
hai rotto la seconda et la prima volta, allhora lassandoli andare tucta dua et tu solamente per tucta dua rotti nota bene della seconda et della prima, tieni alle mani uno in tucto. Et a ciò meglio conphrenda, ti darò qui di sotto lo exemplo di ciaschuno. Et prima venendo rotto la prima volta et non la seconda, se l'amico tuo harà pensato 7, multiplicato per 3 fa 21, partito per 2 ne viene 10 1/2, lassato andare mezo, resta 10 el quale di novo multiplichato per 3 fa 30, partito per 2 ne viene 15. Ecco che la seconda volta non hai rotto. Dimandato quante volte 9 entra in 15, dirà "una" et avanzerà 6, il che in questa dimostratione non si cura quello avanzerà comme nella superiore. Et perhò solo dimanda quante volte 9 entra nella sua somma che in 15 entra una volta solamente. Adunque hai alle mani 4 et, per il rotto che havesti la prima volta, solamente hai 3, posto sopra 4 fa 7. Dunque dirai pensassi nell'animo suo 7 venendo el rotto la seconda volta et non la prima. Exemplo: se l'amico pensassi 6, multiplicato per 3 fa 18 | partito 9v
per 2 ne viene 9 che la prima volta non hai rotto, di novo multiplicha per 3, 9 fa 27 che partito per 2 ne viene 13 et 1/2. Lassa andare el mezo, hai 13 et per questo rotto la seconda volta solamente hai a tener alle mani 2. Dimandando quante volte 9 entra nella sua somma, cioè in 13, dirai una volta che hai 4 et per il rotto che fu la seconda volta, solamente hai 2, posto sopra 4 fa 6 per la somma prese lo amicho. Exemplo: se il rotto viene la prima et la seconda volta, comme se l'amico pigliassi 9, multiplicato per 3 fa 27, partito per 2 ne viene 13 1/2. Lassa andare el 1/2, resta 13, el quale di novo multiplicato per 3 fa 39, partito per 2 ne viene 19 1/2, lascia andare el 1/2, resta 19 che 9 vi entra 2 volte per li quali hai 8 et perché tucte dua le volte hai rotto, hai solamente a tener uno alle mani, posto sopra 8 fa 9, per el numero si pose l'amico in quore hai tucti li exempli di tucti e modi ti può venire rotto. Et nota che mai questa regola non falle, operandola bene. Hor nota uno altro modo.

Demostratio quarta a trovare uno numero intero senza rotto

Una altra legiadra et bellissima regola | anchora serve a trovare uno numero 10r
intero che non vi sia rotto, ma sia intero et sia quanto si voglia. Terrai questo modo, cioè dirai allo amico tuo che 'l numero pensato multiplichi per 3 et

29 dua] K seconda 34 altra] K alta

quello parta per 2. Et nel partire dimanda se vi viene rotto. Venendovi dirai lo facci intero, cioè dica uno più. Et poi dicto partimento, cioè quella metà, multiplichi di novo per 3 et quello ne viene parta per 2. Et s'el partire vi venisse rotto, dirai una altra volta lo facci intero et tu per la seconda volta tieni sempre alle mani 2. Et dirai, facto questo, ti dica quante volte entra 9 nella sua somma, il che saputo, per omni volta che 9 entra nel numero, hai a tenere a mente 4. Et sopra questo posto le unità hai alle mani per li rotti, sarà el numero prese l'amicho tuo. Et simile regola mai non falla. Exemplo: se l'amico pensassi 9, dirai lo multiplichi per 3, farà 27, partilo per 2, ne viene 13 1/2 che hai rotto la prima volta, fallo intero et dirà 14 et tu per questo primo rotto hai alle mani uno. Et poi questo 14 lo multiplichi per 3, farà 42, partito per 2 ne virrà 21 et non hai rotto la seconda volta. Dimanda dunque quante volte entra 9 nella sua somma? Dirà: 10v "dua". | Quello avanza non si cura et per le 2 volte entra nella sua somma, ti denota 8 et uno havevi alle mani per il primo rotto, fa 9 et tanto si pose l'amico. Et se lui pigliassi 10 dirai multiplichi per 3 fa 30, partilo per 2, ne viene 15 et non hai rotto. La prima volta multiplica 15 per 3 fa 45, parti per 1/2 ne viene 22 et 1/2 che harai rotto. La seconda volta dirai lo facci intero, fa 23 et tu per questa seconda volta tieni alle mani 2. Dimanda quante volte entra 9 nella sua somma, dirà 2 che ti denota 8 et havevi alle mani 2 per il rotto e' quali sopra 8 posti fa 10 per il numero si pose l'amico. Ma se lui penserà 7 che multiplicato per 3 fa 21, parti per 2 ne viene 10 1/2 che hai rotto. Fallo intero, sirà 11 et per questo primo rotto tieni alle mani uno, multiplicato 11 per 3 di novo fa 33, partilo per 2 ne viene 16 1/2, fallo intero dirà 17 et per questo secondo rotto tieni alle mani 2 et havevi uno per il primo rotto, fa 3. Di poi dimandato quante volte entra 9 nella sua somma, dirà una per la quale hai 4 et 3, per li rotti hai alle mani fa 11r 7 et tanto dirai, si pose l'amico. Potrà essere ancho|ra che mai non verrà rotto comme se l'amico pensassi 8, multiplicato per 3 fa 24, partito per 2 ne viene 12, el quale di novo multiplicato per 3 fa 36, partito per 2 ne viene 18. Eccho che ne hai rotto nessuno. Adunque in questo dirai che in 18 entra 2 volte 9 che l'amico si ponessi 8 et nota che omni volta che lo amico piglierà numero che partito per 4 non vengha rotto comme 4, 8, 12, 16, 20 et così in infinito mai ti verrà rotto né prima né la seconda volta, ma usciendo di questi numeri sempre o alla prima o alla seconda verrà rotto et questa regola mai ti fallirà. Havete 4 modi da ritrovare uno numero pensato per l'amico intero, cioè senza rotto ei quali, benché tucti si fondino in su una regola, tucti sono variati. Nel conspecto delle persone parranno varii et sono perché, benché sieno in una medesima

14–15 et tanto si pose l'amico. Et se lui pigliassi 10] F et tanto si pose. Et se l'amicho pigliassi 10
24 et havevi uno] F e 1 havevi

regola fundati, tamen omni uno in sé è variabile che l'uno non va per l'ordine dell'altro. Et hora a seguitare avanti diremo così.

Demostratio quinta a trovare uno numero che havessi rotto

Per insino ad hora hai 4 modi a trov|are uno numero intero che non sia rotto, 11v cioè che in sé non habbi parte rotte o divise. Hora mi pare di necessità darti qualche modo da trovare uno numero intero con rotto et rotto o intero senza rotto. La regola fia generale a che numero tu vuoi. Terrai questo modo, cioè dirai che tucta la somma si trova fra interi et rotti, la doppi et poi vi pongha su 5 et poi tucto multiplichi per 5 et poi sopra la somma ponga 10 et poi el tucto multiplichi per 10 et facto che hai queste multiplichationi mutamente di tucta la somma caverai 350, overo farai cavare a llui. El rimanente partirai per 100 et quello ne viene sarà el numero pensato per lo amicho tuo. Et mai ti fallirà questa regola, comme per exemplo dirò. Poniamo che lo amicho habbi preso 8 et 3/4, doppiato fa 17 et 1/2, posto in su 17 1/2 5 fa 22 et 1/2, multiplicato 22 1/2 per 5 fa 112 et 1/2, postovi su 10 fa 122 1/2 multiplicato per 10 questo 122 et 1/2 fa 1225. Dunque caverai di questo 1225 350, resta 875, el quale numero partito per 100 ne viene 8 75/100 che schisati sono 3/4 apuncto. Dunque dirai si ponessi 8 3/4 et | così farai in omni numero che mai ti fallirà. Et se l'amico pigliassi 5 et 1/2, 12r doppiato fa 11, postovi su 5 fa 16, multiplicato per 5 fa 80, postovi su 10 fa 90, multiplicato per 10 fa 900, tractone 350 resta 550, partito per 100 ne viene 5 1/2 per il numero si pose l'amico. Et se l'amico pigliassi numero rotto solo comme 2/3, doppiato fa 1 1/3, postovi su 5 fa 6 1/3, multiplicato per 5 fa 31 2/3, postovi su 10 fa 41 2/3, multiplicato per 10 fa 416 2/3, tractone 350 resta 66 2/3 che, partito per 100, ne viene 2/3 apuncto per il numero si prese. Et quando pensassi numero intero solamente, anchora serve questa regola. Comme se l'amico pensassi 7, doppiato fa 14, postovi su 5 fa 19, multiplicato per 5 fa 95, postovi su 10 fa 105, multiplicato per 10 fa 1050, tractone 350 resta 700 apuncto che, partito per 100, ne viene 7 apuncto per il numero che pensò l'amico. Et questa regola mai ti fallirà o rotto o intero et in tucti e' modi questa regola fia universale.

Demostratio 6 a trovare uno numero quando fussi rotto et sano | et generalmente in qualunche modo 12v

Una altra vera regola serve quando volessi ritrovare uno numero pensato per lo amico tuo in omni qualunche modo si sia, o vuoi intero solo, o rotto solo, o rotto et intero insiemi. Comme la superiore sarà la regola generale, terrai modo seguente, cioè dirai allo amico che di questo suo numero ne facci 2 parti a suo

28–29 ti fallirà] F fallirà 30 et sano] K et sano et sano 35 cioè dirai] K cioè et dirai

modo che non fa caso o interi o rotti, dalli libertà. Et poi dirai che omni una di decte parte multiplichi in sé medesimo et tucto acozzi insiemi, poi dirai che le due parte multiplichi una con l'altra et che questa multiplichatione la doppi et questa doppiata la ponga in sul dicto numero della multiplicatione di ciaschuna in sé, cioè nel quadrato di omni una in sé perché a dire multiplicha uno numero in sé medesimo denota quadrare quello tale numero. Dunque dirai che multiplichi le 2 parti che lui fe', l'una nella altra, et questo multiplicato doppi et poi li pongha in su li quadri delle parte et poi ti dica la somma et tu di 13r questa somma piglierai la sua radice et quella sarà el numero prese l'amicho. | Exemplo: se l'amico havessi preso 8, factone 2 parti comme voi non fa caso rotta o sana, non inporta equali o in equali, dalli la libertà, diremo adunque ne facci 2 parti, cioè 5 et 3. Multiplica omni una in sé, harai 5 via 5 fa 25 et 3 via 3 fa 9, acoza 25 et 9 insiemi fa 34. Hora dico che multiplichi le 2 parte una nell'altra, cioè 3 in 5 fa 15 et questo doppia, fa 30, posto sopra ei dua quadrati, cioè sopra 34, fa 64, la cui radice è 8. Et tanto si pose. Et se pigliasse 7 et dividessi in due parti equali, cioè 3 1/2 per parte, dirai medesimamente che quadri tucta a due le parti et i quadrati giuncha insiemi dove quadra 3 1/2, farà 12 1/4. Et l'altro 3 et 1/2 similmente 12 1/4, giuncti insiemi faranno 24 1/2. Dipoi multiplicha le parti l'una nell'altra faranno 12 1/4, questo doppiato farà 24 1/2, posto sopra 24 1/2 hai prima, farà 49, la cui radice è 7 et tanto dirai che l'amico si fusse posto. Et quando el numero fussi intero et rotto farai così similiter. Comme se havesse preso 6 1/2, diviso in due parte 2 1/2 e l'altra 4 dove multiplica omni una in sé 13v arai | 2 1/2 in sé fa 6 1/4 et 4 in sé fa 16. Acoza insiemi, farà 22 1/4 et poi multiplica 4 via 2 1/2 fa 10, adoppiato fa 20, postovi sopra 22 1/2 farà 42 1/2, la cui radice fia 6 1/2 et tanto prese. Et quando pigliassi 1/2 diviso in 2 parte: 1/4. 1/4 multiplicato omni una in sé fa 1/16 et 1/16 acoza insiemi fa 2/16, multiplicato l'una nell'altra fa 1/16, adoppiato fa 2/16, posto sopra el quadro delle parte in sé, fa 4/16 che è 1/4, la cui radice è 1/2 et questa regola mai fallirà et è bellissima.

Demostratio septima a trovare uno numero per l'amico pensato generalmente

Diremo uno altro modo a trovare uno numero generaliter in tucti i modi, o rotto o sano non porta ad libitum. Farai così: dirai allo amicho che facci del numero dua parte a suo modo o rotte o intere non fa caso. Poi dirai che quadri le dicte parte multiplicando omni una in sé. Et poi questi quadrati multiplichi omni uno per 3, di per sé facta questa tale multiplicatione, dipoi multiplichi

---

2 medesimo] K medesime 3 una con l'altra] F una nell'altra 9 piglierai] F pigliano | quella] F quello 10–12 Exemplo … in sé, harai 5] F *omissis*. In F troviamo espunto: et se l'amico pigliassi el numero con rotto comme 8 et 2/3, factone dua parte, l'una 6 e 1/3 et l'altra dua 1/3 quadra.

l'altre parte ciaschuna, cioè la multiplicatione del quadrato della seconda in tre multiplichi poi via la prima parte et il quadrato della | prima in 3, multiplichi [14r] nella seconda et di poi cuba omni una di dicte parte di per sé et tucto acozerai insiemi. Et dicati la somma tucta della quale tu piglierai la sua radice cuba et quello fia el numero preso. Poniamo l'amico pigliassi 8, factone 2 parte, 3 et 5, quadra ciaschuno, harai 9 et 25. Dico che omni una multiplichi per 3 dove 3 via 9 fa 27 et 3 via 25 fa 75. Ora dico che multiplichi 75 per la prima parte, cioè per 3, farà 225 et 27 via la seconda parte, cioè via 5 farà 135. Accozza insiemi 135 et 225—fanno 360; ora cuba omni una di dicte parte, cioè 3 et 5, cuba tre farà 27 et cuba 5 et fia 125. Hora accozza insiemi 360, 27, 125—faranno 512 la cui radice cuba fia 8—el numero prese l'amicho. Et se l'amico pigliassi numero con rotto comme 8 et 2/3, factone dua parti, una 6 et 1/3, l'altra 2 et 1/3, faranno 512 la cui radice cuba fia 8—el numero prese l'amicho. Et se l'amico pigliassi numero con rotto comme 8 et 2/3, factone dua parti, una 6 et 1/3, l'altra 2 et 1/3, quadra 6 et 1/3 farà 361/9, multiplicato per 3 farà 1083/9, multiplicato per 2 et 1/3, cioè nell'altra parte, farà 7581/27, cubato dicta 6 et 1/3 farà 6859/27. Quadra l'altra parte, cioè 2 et 1/3, farà 49/9, multiplicata per 3, farà 147/9, multiplicata via l'altra parte, cioè via 6 1/3 farà 2793/27 cubato dicto 2 et 1/3 farà 343/27; | [14v] dove acozato tucto insiemi, cioè 6859/27 et 7581/27 et 2793/27 et 343/27 harai una somma di 17576/27, la cui radice cuba fia 8 et 2/3, cioè el numero per lo amicho preso. Et così tornerà sempre quando l'amico pensassi uno numero rotto solo et partissilo in più parte rotte. Multiplicandolo per i sopradecti modi et cubandolo, sempre ti verrà benissimo ché mai ti fallirà, ma bisogna che chi fa tale numero a mente sia experto nel travagliare di rotti ché, quando le parti sono rotte, con grandissima dificultà si farà a mente che con fatica si farà con la penna.

Demostratio octava, uno numero in qualunche modo a trovare

Per uno altro modo dissimile a superiori potrai venire alla notitia di uno numero rotto o intero comme si sia non importa. Farai a questo modo: prima dirai allo amicho che del numero pensato ne facci 2 parte equali, cioè lo divida per mezo apuncto. Et poi dirai che a una di queste parte agiungha uno numero a suo modo: o 100 o 1000 o uno non inporta, o rotto o intero a suo comodo, ma dicati quanto vi agiugne che se non lo sapessi tu, non potresti pervenire alla notitia dicta. Perhò farai dica quanto | vi pone su in una delle dicte parte. Poi [15r] dirai che multiplichi una di dicte parte in sé qual vole, o quella in su la quale

3 cuba] F scriba | acozerai] F acorrai 25 sono] K sone, F sonno 28 modo dissimile] F modo medesimo simile 29 importa] K porta 34 farai dica] F farai che dica | su in] K su in su

e' pose el numero o l'altra, dalli libertà, et questo quadrato tengha. Dipoi dirai multiplichi l'altra parte, cioè l'una parte nell'altra, cioè la coniuncta nella semplice et quello fa, ponga sopra el quadrato ha alle mani. Et tucta questa somma partirai per il coniuncto di decte 2 parti insiemi. Sempre ne virrà la parte facesti quadrare. Et perhò bisogna quando lui quadra una parte che tu sappi quale lui quadra, se quadra la composta o la semplice. Et quando lui partirà nel sopradicto modo, sempre ne verrà la parte che l'amico quadrò et in questo modo harai notitia del numero prese lo amico. Eccoti lo exemplo più chiaro: diremo l'amico haverà preso 10, partito in 2 parti equali farà 5 per parte. Hora dirai che in su una di decte parte ponga uno numero a suo modo, non importa, dalli la libertà. Diremo vi ponessi 6, fa 11, hai dua parte: una 11, l'altra 5 a te incognita, ma sai bene che in su una parte vi pose su 6. Hora dirai che multiplichi una di dicte parte in sé qual vole, ma ti dica qual lui multiplica o la composta che è 11
15v o la semplice che è 5. Diremo multiplica|ssi la composta in sé fa 121 et questa serbi. Dipo' dirai multiplichi una nell'altra, cioè 11 via 5 farà 55 et questo porrai sopra 121, farà 176 et questo parti per el congiuncto delle 2 parti insiemi, cioè 11 et 5 fa 16. Dunque parti 176 per 16, ne virrà 11 per el numero quadro et tu sai che lui quadrò el numero composto in sul quale sai che pose 6, dunque leva 6 di 11, resta 5 per una delle parte equali che lui fece del suo numero; dunque dirai si pose in core 2 volte 5 quale è 10. Et se lui havessi quadrato la parte semplice, cioè 5 farà 25 et poi multiplicato l'una nell'altra, cioè 11 via 5 fa 55, posto sopra el quadrato delle parti farà 80, el quale partito per el congiuncto di 11 et 5, cioè per 16 ne viene 5 per la parte quadrò. Et tu sai che lui quadrò la parte semplice, adunque dirai si ponessi in cuore due volte 5 fa 10 et così farai sempre. Potrevi comme nella precedente dare li exempli de' rotti, cioè se lui pigliassi numero rotto o se lo dividesse in parte rotte, ma troppo saria prolisso et darìa fastidio
16r a chi legge et a me medesimo ba|sta solum che questa et tucte le regole darò in questa opera sono tucte perfecte et provate che sono verificate. Chi vorrà, potrà fare lui prova di tucto et troverrà essere la verità di quello che è scripto a starne sempre alla riprova di chi volessi contradire et basti hora alla opera nostra, diremo così.

Demostratio nona a trovare generaliter omni modo

Se volessi per uno brieve modo trovare uno numero per lo amicho postosi in cuore, terrai questo leggiadro modo bello et breve. Dirai che pigli uno numero a suo modo, rotto o intero non inporta, et di questo numero, dirai, ne facci quante parte a llui parrà: o dua o 20 o 100 non inporta, et intere o rotte a suo modo che a

12 ma sai] F assai

te non inporta. Et, perhò, darali libertà che quando ne dai libertà di fare quello vole, pare più bello et più stupente el giocho, siché dirai, in quel modo lui vole facci. Dirai, dunque, ne facci quante parte vole et poi dirai che ciaschuna di decte parte che lui ha facte che le multiplichi omni una di per sé nel numero prese nell'animo et tucte decte multiplicationi agiuncte insiemi, dirai te dica la somma che fanno tucte giuncte insiemi. | Et tu di questa somma piglierai la 16^{v} radice et quella sarà el numero prese l'amico apuncto. Diciamo che lo amico havessi preso 10, factone 3 parti: una dua, l'altra 3, l'altra 5. Hai 3 parte: 2, 3, 5. Dirai che multiplichi ciaschuna di per sé nel numero che prese. Dunque lui multiplicarà 3 via 10 fa 30 et poi multiplicarà 2 via 10 fa 20, posto sopra 30 fa 50, et poi multiplicarai 5 via 10 fa 50, posto sopra 50 fa 100, la cui radice è 10. Et tanto dirai, prese lo amicho. Et se llui havesi preso 14, diviso in 4 parte: 2, 3, 4, 5, multiplica 2 via 14 fa 28, et 3 via 14 fa 42, et 4 via 14 fa 56, et 5 via 14 fa 70, dove acozato tucte le 4 somme multiplicate insiemi, harai una somma di 196, la cui radice fia 14. Et tanto prese l'amico. Et così ti verrà quando l'amico piglierà numero rotto et divida in più parti rotti. Se bene lo dividessi in mille parti, non ti fallirà mai comme tu medesimo potrai experimentando provare.

Demostratio 10a a trovare uno numero quomodocunque sit

Nota una altra bella regola che per virtù del | numero potrai pervenire alla noti- 17^{r} tia di qualunché numero si voglia, intero o rotto, al libito dello amicho, non fa caso. Terrai questo modo, cioè dirai che del numero pensato ne facci 2 parti a suo modo, equali o inequali che non porta et poi di' che adoppi el numero prese et questo doppiamento multiplichi via una parte, qual vole non fa caso, et sopra dicta multiplicatione ponga el quadrato della l'altra parte, cioè la multiplicatione dell'altra parte in sé medesima. Questa somma tucta caverai, overo farai cavare el quadrato della parte multiplicato el doppio del numero prese et poi farati dire el rimanente del quale piglierai la radice, la quale sarà el numero prese l'amico tuo. Pognamo che l'amico pigliassi 10, factone 2 parte: 3 et 7. Prima dirai che adoppi el numero prese, cioè 10, farà 20 et questo multiplichi via una delle prenominate parte: o per 3 o per 7, non porta. Diciamo lo multiplichi per 7, fa 140, et postovi su 9, cioè el quadrato dell'altra parte, farà 149. Hora dirai ne cavi el quadrato di 7 in sé che fa 49, cavato di 49 resta 100 apuncto, la cui radice è 10. Et tanto dirai, prese l'amico. Et se lui, comme è dicto, facessi di | 10 2 parte, 17^{v} comme di sopra, et adopiasse 10, fa 20, et questo 20 dove lo multiplicò per 7 lo multiplicassi per 3, ad omni modo tornerà comme se multiplicassi 20 per 3, farà 60, et poi 7 in sé farà 49, posto sopra 60 farà 109, cavane del quadrato della parte per la quale multiplicò el numero prese doppio che ne hai a cavare el quadrato di 3, cioè 3 via 3 fa 9, di 109 resta 100 apuncto et la sua radice è 10. Et tanto prese l'amico tuo. Et così farai sempre che ti tornerà bene quando l'amicho piglierà el rotto etc. comme provando potrai trovare ché bisogna che tu, lectore, duri uno poco di faticha ancora tu perché ti do el modo bello et legiadro. Ma el travagliare di rotti è fastidioso in tucte le cose et perhò lasseremo lo exemplo di rotti. A me basta dire cose che habbino bono fundamento et che non si possino riprovare false che so ché delli apuntatori non manchano, ché chi dice una cosa chi dice l'altra, et perhò parlerò giustificato et chi vol dire dica.

Demostratio XIa a trovare uno numero. Regola generale

Abbiamo una altra regola belissima a ritrovare uno numero a tuo modo, o rotto
18r o intero, a tucti | i modi comme ti parrà, la quale fia questa. Videlicet dirai
allo amicho che del numero preso ne facci 2 parti equali. Poi dirai che sopra el numero preso vi ponga su uno numero a suo modo o a tuo, non inporta. Et poi dirai che tucta la somma, cioè el numero preso con quello vi pose su tucto insiemi, multiplichi per el numero vi pose su et questa somma serbi. Et poi dirai che in su una di decte parte equali ponga el medesimo numero pose in tucta la somma et questo tucto insiemi poi multiplichi in sé et di tucta la somma farà cavii el numero serbato, cioè la prima multiplicatione. Et quello resta, ti dica. Et tu allora piglierai la radice del dicto numero et quella sarà la metà apuncto di quanto prese l'amico. Et mai ti fallirà comme per exemplo potrai vedere. Hora eccoti lo exemplo che più chiaro mostra che nulla. Pognamo che l'amicho pigliassi 10, diviso in 2 parti equali, harai 5 et 5. Hora ponga in sul numero prese, cioè in su 10, uno numero a suo modo. Diremo vi ponessi 6, farà 16 et questo multiplicato per il numero agiuncto, cioè per 6 farà 96 et questa serba. Hora dico che in su una di decte parte ponga el numero pose in su el tucto,
18v cioè questo 6 ponga in su 5, fa | 11 et questo multiplichi in sé, fa 121 et questo
dico che cavi la prima multiplicatione che havevi, cioè 96 di 121, restarà 25, del quale piglierai la radice che 5 che fia la metà del numero prese. Dunque dirai che l'amico pigliassi 2 via 5, fa 10, et tanto prese. Et così farai sempre et mai ti fallirà.

Demostratio XIIa a trovare uno numero generale

Potrassi per uno altro legiadro modo venire alla notitia di uno numero pensato, intero o rotto, a tuo modo et fassi in questa forma: prima dirai che del numero che lui ha preso ne facci 2 parte a suo modo, equali o inequali non inporta, et dirai che ti dica una di dicte parte, qual vole, et l'altra serbi in sé. Hora dirai che la parte a te incognita ponga sopra el numero prese di principio et poi tucta la somma multiplichi in sé medesimo et poi tu multiplica la parte a te nota in te medesimo. Et dirai: cavi la tua somma della sua. Et poi di quello li rimane, faralo partire per 4 et quello ne viene, dirai, lo parta per el numero prese di principio
19r et verrane l'altra parte a te incognita et tu l'acozzerai con la parte a te | nota et

harai la somma prese l'amico. Comme se epso havesse preso 10 diviso in 2 parte: 6 et 4. Poniamo che a te sia nota la parte 4, dove dirai allo amico che sopra el numero prese di principio ponga l'altra parte, cioè sopra 10 che prese ponga 6, cioè l'altra parte, farà 16, et questo multiplichi in sé medesimo, cioè 16 via 16 fa 256 et di questo farai cavare el quadrato della parte a te nota, cioè di 4, dove multiplicato 4 in sé fa 16, cavato di 256 resterà 240, et questo 240 farai partire per 4 che ne verrà 60 apuncto. Dirai, dunque, che parta 60 per el numero prese che llo partirà per 10, ne verrà 6. Per la parte a te incognita, dove acozerai 6 et 4, farà 10 et tanto, dirai, pigliassi l'amico di principio. Et se fussi stato noto 6 et non 4 et dectoli che sopra el numero pensato pongha l'altra parte, harà posto 4 sopra 10 farà 14 et questo 14 multiplica in sé, farà 196, cava la multiplicatione della parte a te nota in sé medesima, cioè 6 via 6 fa 36, cavato di 196 resta 160, et questo, partito per 4, ne viene 40, dove, partito 40 per el numero prese di principio, cioè per 10, ne verrà 4 | per la parte a te incognita. Eccho la regola riescha 19v sempre, quando bene pensassi rotti sempre te riuscirà, non inporta, operando bene.

Demostratio XIIIa a trovare uno numero quomodocunque sit

Troverai anchora uno numero per lo amico, excogitato in qualunche modo si sia per lo amico pensato. Farai in questo modo, sia el numero di che qualità si vole. Prima dirai allo amico del suo numero che ne facci due parte di che qualità si vole, o equale o inequale, non porta, et poi dirai che omni una di queste parte multiplichi in sé, cioè che le quadri. Et questi dua quadrati ponga l'uno sopra l'altro et tenga a mente. Poi dirai che pigli la differentia di una di decte parte alla metà del numero prese di principio et questa differentia multiplichi in sé medesima. Et quello che farà, cavi della metà della somma havevi alle mani, cioè della metà della somma di 2 quadrati sopradecti et quello li resta, farai ti dica a tte, del quale numero tu piglierai la radice et quella sarà la metà del numero che lui prese et mai ti falla. Exemplo: se l'amico pigliassi 10, factone 2 parte quomodocumque | non porta, l'una 7 et l'altra 3, dirai che ciaschuna di 20r loro multiplichi in sé, cioè 7 via 7 fa 49, et 3 via 3 fa 9, achoza insiemi 49 et 9 fa 58, et questo serba. Hora piglia la differentia che è della metà del numero pensato, cioè da 5 di una di dicte parte qual voi: o a 7 o a 3, non inporta, che troverrai che da 7 a 5 ci è 2, medesimamente da 3 a 5 vi è 2. Dunque multiplica 2 via 2, cioè la differentia che vi è in sé, farà 4, et dirai che pigli la metà di dua quadrati haveva alle mani et di questo cavi el quadrato della differentia sopradicta, dove in questa harai a cavare 4 della metà di 58, cioè di 29, resterà 25 et tu di questo

12 cioè] K ciò 26 cioè] K ciò

25 piglia la radice che è 5 per la metà del numero prese l'amico. Dunque dirai 2 via 5 fa 10 et tanto dirai che pigliasse l'amico tuo in cuore et così fa sempre.

Demostratio XIIIIa a trovare uno numero generale

Potrai trovare un numero per lo amico, preso in qualunché modo comme di sopra, rotto o intero, non dà noia, facendo in questo modo: prima dirai allo
20v amicho che el suo numero che ha preso vi ponga su altro numero | a suo modo
comme li piace, o rotto o intero, non dà noia, et poi tucta questa somma multiplichi in sé medesima et serbi. Et po' dirai che pigli el numero che lui pose sopra el suo pensato, et questo medesimamente multiplichi in sé et ponghalo sopra el quadrato del numero prese et di questo pigli la metà et tengha a mente questa somma. Et poi dirai che pigli la metà del numero che lui prese da principio et in su questa metà ponga el medesimo numero, pose in su tucta la somma et tucto questo multiplicato in sé medesima. Et questa somma cavi della somma hai alle mani. Et quello rimane farai ti dica, del qual numero piglerai la radice, la qual fia la metà del numero prese l'amico. Et così farai sempre. Eccoti lo exemplo più chiaro: habbi preso l'amico 10 et sopra questo 10 ponga altri 10 fa 20. Hora dico che questo 20 multiplichi in sé medesimo: 20 via 20: 400. Hora dico: "pigli el numero, ponesti sopra el numero preso che è 10 et questo multiplichi in sé medesimo, cioè 10 via 10 fa 100". Et questo poni sopra 400 farà 500, del quale getta via la metà, ti resterà 250 et questo serba in mente. Hora piglia la
21r metà del numero pigliasti, | cioè 5 et sopra questo poni el numero ponesti in
su tucta la somma, cioè 10, farà 15. Et questo multiplica in sé medesima, cioè 15 via 15 farà 225, et questo cava del numero havevi alle mani, cioè di 250, resterà 25. Et farai ti dica questa ultima somma che sino a qui non bisogna ti dica nulla hora di questa ultima somma et tu piglierai la radice che la radice di 25 fia 5 per la metà del numero prese lo amico. Tu dunque dirai: piglassi 2 via 5 fa 10 et tanto prese di principio. Et così facendo, sempre ti riuscirà. Pigli rotti a suo modo non fa nulla perché la regola è generale in tucti i modi che lui piglerà o vorrà, comme è dicto, o rotti o sani. Farai tu la prova s'el ti dilecta fare provare di rotti et troverai essere la verità di quello ti ho mostrato in tucti i modi.

Demostratio XV a trovare uno numero generaliter

Benché io ti habbia mostro di sopra nelle predicte dimostrationi et regole molti belli et varii modi, diversi et differenti in opera et in facti, con quali potrai
21v venire alla vera et perfecta notitia di tucti i numeri | si potessi nell'animo di
uno pigliare, anchora te ne voglio dare uno, el quale, benché sia noto universalmente a molti, niente di mancho darò via et modo da coprire la sua facilità. Et in facto chi vorrà quello operare nel modo li darò, non lo farà senza difficultà. Questo modo lo inparai quando nelli teneri anni andavo alla scola dove

vanno i piccoli fanciulli nelli principii delli loro studi. Et dicevasi così: “pigla uno numero in core”, el quale preso, diciavamo: “partilo per 3 et dimmi quello ne viene”. Dove saputo quello, ne veniva a partire per 3. Multiplicavamo quel numero per 3 et havemmo el numero intero si prese. Comme se lui havessi preso 12 partito per 3 ne viene 4, dove multiplicato 4 per 3 farà 12 apuncto. In questo modo non ci mettendo altra operatione, so qui direbbe el mio gentile lettore: questo non achadeva perché me lo sapevo et troppo è la sua facilità. Dove ti voglio dare qualche modo brieve da coprire questa sua facilità, benché so che apresso li acutissimi ingegni non si mostra el bianco per lo nero. Pure non ti spiaccia di leggere che forse | non ti parrà da sprezarla et omni uno non 22r
la intenderà così facilmente. Dirai allo amico tuo che pigli uno numero a suo modo et di che sorte et qualità si vole, et poi dirai che lo multiplichi a suo modo o per 3, o per 4, o per 6, o per 100, non ti dà noia, et multiplichi una o 2 o 4 o 6 volte quante a llui pare, che tu li dai piena libertà. Ma bene ti dica che quando lo multiplica per 4 o per 6 o per 10 et quante volte et a te bisogna tenere a mente tucte queste sue multiplicationi, cioè quando lui lo multiplica per 2 et poi per 3, etc. Hora quando lui si vole fermare che lui non vorrà più multiplicare, dirai che ti dica tucta la somma che lui ha et tu bisogna che vadi partendo quella sua somma per tucti li numeri che lui l’ha multiplicata et verrai alla notitia del numero prese lo amico tuo. Comme se lo amico havessi preso 8, hora tu li dai la libertà. Lui prima dirà: “io lo multiplico per 2 farà 16”. Et poi dirà: “io lo multiplico per 5 farà 80”. Et poi dirà: “io lo multiplico per 4 farà 320” et poi dirà: “io lo multiplico per 3 farà 960. Hora diciamo che lui si fermi et dica non volere più multiplicare. Farai ti dica la somma che lui ha alle mani et tu va ripartendo per li numeri che lui ha multiplicati | che l’hai a partire per 2, per 5, per 4 et 22v
per 3, dove per fare più presto che uno pratico nel partire potrà fare così: dire 2 via 5 fa 10 et poi questo 10 multiplica per l’altra figura, cioè per 4 farà 40 et questo 40 multiplica per 3 farà 120. Partirai tucta la sua somma per 120, ne virrà 8 apuncto per il numero prese. Ma se tu non te ricordassi partire per sì grande numero, puoi partire numero per numero di per sé, dove parti 960 per 5, che non fa caso da qual numero ti cominci a partire pure che tu lo parta per tucti i numeri che lui multiplicò el suo o prima o poi, non dà noia. Dunque parti 960 per 5, ne virrà 192. Parti di poi 192 per 4, ne verrà 48. Partirai poi 48 per 3, ne virrà 16. Partirai poi 16 per 2, ne virrà 8. Eccoti el modo. Poteresti ancora partire in uno altro modo, cioè qui tu hai quattro partitori. Partili a 2 a 2, cioè partir la somma sua prima per 2 via 5 fa 10, et per 3 via 4 fa 12, dove parti 960 per

13–14 o per 100, non ti dà noia, et multiplichi una o 2 o 4 o 6 volte quante a llui pare] F o per 10, tante volte quante a llui pare

10, ne virrà 96 et poi parti 96 per 12, ne virrà 8 apuncto. Hora tu piglierai quello modo che più ti satisfarà et che ti fia più facile. Potrai anchora coprire la facilità 23r di questa regola | in uno altro modo. Comme a dire: quando lui ha fermo che lui non voglia multiplicare più et tu a ciò non conphrenda l'ordine tuo, dirai che ne cavi una somma et tu tieni a mente quello che li fai cavare, verbi gratia. Della sua somma ne cavi 32 et poi dirai ne cavi 18 et hai in tucto facto cavare 50 che sono 2 volte 25. Hora dirai: "ne cavi 13", et poi dirai": ne cavi 12" et già hai 3 25. Et poi puoi dire: "cavane 130", et poi: "levane 20" et hai 150 che sono 6 25 et tre havevi facto cavare prima, fanno 9. Hora dirai: "ne cavi 25" et hai 10 volte 25 che ne hai facto cavare 250. Hora dirai ti dica el resto che se lui harà della multiplichatione sopradicta 960, cavatone 250 resterà 710. Farai te lo dica et tu porrai in su 710 la somma ne hai facta cavare, cioè 250 et andarai ripartendo comme di sopra. Et torneratti bene comme di sopra, potrai tu in questa regola fundare varie et legiadre regole et modi diversi che saria prolisso lo extendermi in simile materia, alli homini intelligenti basta uno minimo cenno.

23v **Demostratio XVIa a trovare uno numero pensato per forza et virtù di unità et fare con epsa unità mirabil forza**

Quanta forza et virtù habbi in sé la sanctissima unità per li sui mirabili effecti, si vede et dimostra dove in questo nuovo capitulo intendo dare notitia particulare di parte, di sua mirabilissima forza. Et veramente, lectore mio gentile, questo capitulo inparalo a mente perché chi quello havessi bene a memoria, li farebbe uno honore mirabilissimo che se uno durassi tucto dì a multiplichare uno numero solamente con la unità tua, saprai dirli sempre a che termine si trova, cose invero laudabile. Et questo giocho lo inparai dalla felice anima di madonna Alexandra, figliola di Messer Bartolomeo Scala, quando nella mia tenera età da llei inparai quello pocho io so delle scientie dicte humanità. Benché mi reputo niente sapere, tamen a quelli benedicti ossa sono et sarò sempre obligatissimo, alla cui Idio doni requie. La quale Madonna Alexandra per le sue innumerabili virtù meritò essere coronata dell'arbore sì grata ad 24r Apolline che sia il vero | in nel vulgo anchora l'è chiamata et nominata poetessa. Da llei dunque inparai questo digno giuocho et a llei lo atribuischo che veramente è uno bello secreto comme leggendo vederai. So bene che qualche uno dirà: "costui doveva havere captivi vicini". A questo non rispondo perché l'opera risponderà lei et anchora non voglio così rispondere ad omni uno che

---

5 verbi gratia] F dirai verbi gratia 14 lo] F per lo 21 quello havessi] F questo capitolo havessi 24 felice anima] K felica aima 25–26 nella mia tenera età] F nelli miei puerili anni 27 tamen] F pure 28 Idio] F Dio 30 Apolline] F Apollo

harei troppa facenda. Qui bisogna badare ad altro che a rispondere a chi dice. Attendendo all'opera nostra, diremo così: se uno tuo amico o parente pigliassi uno numero rotto o intero, non inporta, a suo modo, et tu, per virtù di unità, volessi ritrovarlo, dirali così: che faccia del numero suo quello vole, cioè che lo multiplichi per quello li piace. Dicati sempre per quanto lo multiplicha et tu piglia la unità et quello che lui farà della sua somma, quello farai tu della tua unità. Et quando lui non vorrà più multiplichare, tu li dirai che parta la sua somma per la somma hai tu alle mani et verranne el numero che lui prese. Overo dirai che parta el numero che lui si trova, cioè l'ultima somma per quello prese et verranne la tua somma. Hora nota lo | exemplo più chiaro. Poniamo 24v
che lo amicho o compagno pigliassi 6 et dicessi: io multiplicho per 5, farà 30. Et tu piglia la tua unità et multiplichala per 5 farà 5. Poi dirai che l'amico multiplichassi per 6 el suo 30, farà 6 via 30: 180. Et tu hai a multiplichare 5 via 6 fa 30. Hora dirai che lui lo multiplichassi per 4, cioè 4 via 180 farà 720 et tu multiplicha el tuo 30 per 4, farà 120. Hora dirai che lui multiplichassi el suo numero per 3 fa 2160 et tu multiplicha el tuo 120 per 3 farà 360. Hora diremo che lui multiplichassi el suo numero per 2 fa 4320 et tu harai alle mani 720. Hora dirà non volere ire più innanzi et tu dirai che parta el suo numero per il tuo, cioè: 4320 suo per 720 tuo, ne verrà 6 che apuncto per il numero prese. Overo dirai che parta el suo numero per 6, cioè per quello prese, ne verrà apuncto el tuo numero, cioè 720. Ma se lui dicessi: "voglio questo numero mio multiplichare et partire et giugnere et cavare a mio modo che solo el multiplichare lo saprei fare anchora io", et tu rispondi che a questo se' contento, ma nota che el multiplichare et partire del tuo et suo numero, non porta, | multiplichilo et partilo 25r
per quello vole, pure che lo dica a te. Ma el cavare et giugnere vole essere proportionato, cioè non dica: "io ne cavo 10 o 20 o 30" che non ti tornerà bene. Et così dirà: "io vi pongo su 10", etc. Ma bisogna dire: "io vi pongo su el 1/4 della mia somma, io vi pongo el 1/3, 1/8". Overo dica: "io vi pongo su una volta o 2, o 3 el numero che io presi et tu, similmente, porrai tante volte quante lui porrà el suo et porrai la tua unità". Orsù, diciamo che l'amico tuo habbi preso 4 et questo multiplichi per 8, farà 32 et tu piglia la tua unità et multiplicala per 8 farà 8. Hora lui dirà: multiplici el suo 32 per 5, fa 160 et rimultiplicherai el tuo 8 per 5, farà 40. Hora li dica: io la voglio partire per 4 et partito 160 per 4 ne virrà 40. Et tu partirai la tua somma per 4, cioè per 4, ne virrà 10. Nota che se lui dicessi: "io voglio partire la mia somma per lo numero presi", a questo dirai che non vuoi perché non riuscirebbe la tua regola, che la tua unità è immobile, che

13 qui] F farà 24 multiplichilo] F multiplica 26 dica] F tolgha 33 li] K di 34 cioè per 4] F *omissis*

25v multiplica uno numero per la unità mai si | move se bene lo multiplicassi 1000 volte. Et così parti uno numero per la unità, sempre sirà immobile. Et perhò a te non serve che lui dica: “io parto la mia per quello che io presi, cioè per el numero presi”. Ma bisogna ci dica: “io parto la mia somma ultima per 4 o per 6 o per 8”, etc. Et così non ti servirebbe se lui dicessi: “io multiplico el mio numero per lo numero presi” ché, comme è dicto, la tua unità è immobile in sé medesima et mai ti moveresti del numero ti trovassi. Perhò è di necessità lui ti dica quando la multiplica per quanto numero, specificandolo, etc. Siché, tornando al sopradecto caso, lui ha alle mani 40 et tu hai 10. Hora diremo che lui dica: “io ci voglio porre su in questo mio numero el 1/5 del mio producto ultimo che io mi trovo che è el 1/5 di 40 si è 8”. Et tu porrai el 1/5 di 10 in sul tuo che harai 12 et lui 48. Hora lui dirà: “io ci voglio porre su el 1/4 dell’ultimo producto che el 1/4 di 48: 12, farà 60”, et tu porrai 3 che è 1/4 del tuo, farà 15. Et nota che se lui dicessi: “io
26r voglio porre in sul mio numero 10”. A questo rispondi | che non ti serva perché la regola non riuscirebbe, ché tu non sai 10 che parti sia del suo numero et non sapresti che proportione havessi col tuo numero. Dunque dirai che quando lui ha a porre o a levare del suo numero che ti dica solamente: “io pongo el 1/4, el 1/3, el 1/5” et così a scemare che altrimente non ti servirebbe la regola. Hora torniamo al caso nostro. Lui ha alle mani 60 et tu 15 et l’uno non sa dell’altro. Hora diciamo che lui dica: “io voglio porre in sul mio numero 5 volte la somma che io presi” che sarà 4 via 5 fa 20. Et tu dirai essere contento et porrai in su l’altro numero, cioè in su 15 5 volte la tua unità. Harai 20 et lui harà 80. Hora lui dirà: “io voglio multiplicare el mio numero per 3”. Farà 240 et tu harai 60. Hora se lui non volessi più ire innanzi et dicessi: “io mi fermo qui”, allora dirai parta el suo ultimo producto per 60, che ne virrà el numero che lui prese, overo parta el suo ultimo producto che lui si trova per el numero prese, ne virrà 60
26v apuncto che sempre ti tornerà. Ma tu in questo caso potrai | fare una altra gentileza bella, comme dire allo amico: “domandi che numero tu vuoi che io ti facci venire alle mani”. Diciamo che lui dica: “fammi venire 100”. Dove dirai: “piglia uno numero a tuo modo et multiplicalo et partilo et crescilo et scemalo a tuo modo et comme tu non vuoi più multiplicare, ti ferma et io ti farò venire alle mani 100”. Dove, per non essere più fastidioso et lungho che si bisogni, piglieremo el caso superiore et diremo che l’amico habbi alle mani 240 et tu habbi alle mani 60 et che l’amico habbi preso di principio 4 et che lui si fermi qui et dica volere che li provenga alle mani 100 che invero sarà caso mirabile apresso a

---

4 ci dica] F dicha 6 tua] K mia 11 che harai] F e arai 13 farà] F che fa 16 sapresti] F sapessi 17 ha a porre] F a porre 18 hora] F hora nota 23 farà] F harà 24 più ire] F andare più 30 et scemalo] F *omissis* 35 li] F gle

molti. A te bisogna tenere questo ordine, cioè che quello numero che l'amico ti dice voli li facci venire alle mani, che tu habbi alle mani tu quello tale numero. Et non lo havendo alle mani bisogna operi in modo che ti venga alle mani, di che ti darò breve modo. Questo caso sopradecto che lui ha alle mani 240 et tu hai 60 et vorresti havere alle mani 100. Farai così: parti 100 per 60, ne virrà 1 2/3. Ho|ra dico che multiplichi 60 tuo per 1 2/3 farà 100 apuncto. Ecco che hai [27r] alle mani 100: numero da te desiderato. Et poi dirai allo amico che multiplichi el suo ultimo numero per 1 2/3 che multiplicerà 240 per 1 2/3, fa 400 apuncto. Hora dirali: parta el suo ultimo producto per lo numero che lui prese di principio, cioè per 4 che lui prese di principio, li verrà 100. Comme parti 400, che è ultimo producto, per 4, ne viene 100 apuncto. Et per uno altro modo più facile potrai fare el medesimo, in questo caso del fare venire 100 alle mani tua perché basta che vengha a te alle mani el numero chiestoti che comme l'hai tu è facile farlo venire a llui. Comme in questo caso sopradecto che l'amico o conpagno tuo ha alle mani 240 et tu hai alle mani 60 et vorresti havere 100. Potrai fare anchor così. Di': "io ho 60; per ire sino in 100 mi mancha 40". Dico che guardi 40 che parte è di 60 troverrai essere e' 2/3. Hora dirai allo amico che giunga alla sua somma ultima e' 2/3 di decta somma et tu porrai e' 2/3 in su la tua somma et harai 100 apuncto. | Et lui porrà in sul suo 240 e' 2/3 di dicto 240 che sarà [27v] 160 che farà tucto 400 apuncto. Dove partito 400 per el numero prese di principio comme è dicto, cioè per 4, ne verrà apuncto 100, cioè la tua somma. Et se lui volessi sapere quello prese, dirai parta la sua somma ultima per la tua, cioè 400 suo per el tuo 100 et verralli 4 per el numero che prese. Et mai ti fallirà questa regola. Et nota che comme ho dicto di sopra quando fai questo giocho, ti bisogna havere una grandissima advertentia al partire et multiplicare, perché se alla prima multiplicatione lui ti dicessi: "io voglio quadrare el mio numero, cioè multiplicarlo in sé", non ti servirebbe la regola perché, comme è decto, la unità in sé medesima è immobile. Et perhó, comme di sopra ho decto, farai quando tu multiplicerai et partirai, ti specifichi el numero et quando tu poni o scemi che tu scemi proportionato comme apieno ho dicto. Potrai fare questo giocho mirabile in questa forma che si facci dare uno numero da uno altro suo amico et che di quello tale numero li farai venire alle mani che | numero lui vorrà et [28r] faccine lui quello vole: multiplichi, parta, giungha et levi comme di sopra. Et tu dirai che solamente una volta o dua vuoi agiungere o scemare a tuo modo. Et quando lui è fermo che non vole fare più operatione del suo numero et tu dimanda colui che li detti el numero quanto vole che li facci venire alle mani.

12 fare el] F fare per el 13 chiestoti] F chiesto 17 parte è] F è parte | giunga] F agiungha
18 su la] F sulla 23 che prese] F che llui prese 34 agiungere] K ogiognere

Et guarda se tu hai tu quello tale alle mani. Havendolo non ti bisogna fare altro se non farai partire l'ultimo suo producto per lo numero che li dette lo amico et harallo. Ma se tu non lo havessi alle mani, bisogna operi in modo che ti vengha comme di sopra te ho dato el modo. Ma perché te ho dato el modo, quando tu ti trovassi 60 et lui 240 et havessi preso 4, hai a giugnere 1 2/3 comme di sopra. Ma quando tu ti trovassi 120 et lui 480 et lo amico volessi che tu li facessi venire alle mani 50 apuncto, terrai questo breve modo. Dirai: io vorrei havere alle mani 50 et io ho 120. Dove dirai allo amico getti via la metà del suo numero et harà 480, 28v gettando via la metà, resteralli 240. Et tu fa el simile, gettando via | la tua metà di 120, ti resta 60. Hora dirai: "io vorrei 50 alle mani et ho 60". Dove dirai: "io ho più 10 che non vorrai". In questo caso bisogna vegga 10 che per te è di 60 che è 1/6. Hora gitterai via della tua somma 1/6. Rimarratti 50 comme voi et dirai allo amico che getti via 1/6 della sua somma, cioè di 240 che il 1/6 de 240 è 40. Et gettato via di 240 resta 200 apuncto, dove dirai che parta el suo ultimo producto per el numero li dette l'amico, ne verrà 50 apuncto, comme ti ha chiesto l'amico, ché partito 200 per 4 ne viene 50. Hotti dato di sopra el modo facile, ma el bono abachista lo potrà fare la prima volta che li torni alle mani 50. A chi non è bene pratico, potrà fare el modo superiore. Ma volendolo fare in uno tracto dirai così: "io ho 120 et vorrei 50. Dunque ho più 70 che 'l mio bisogno". Dove dirà: "70", che parte è di 120 vedrai essere 7/12. Dove dirai allo amico che del suo 480 cavi e' 7/12, gli resterà 200 apuncto. Et tu cava di 120 e' 7/12, ti resterà 50. 29r Ecco che hai tucti i modi da potere | fare el gioco bello et vago nel conspecto di tucti mirabile. Et se in questa dimostratione mi sono disteso a llungho, ti dico che non si pò dire le cose sottile in breve parolle, ché a volere che ciaschuno sia capace di questa regola, bisogna dire aperto modo, si ché habbimi lectore a scusa. Et leggiolo tucto che non passerà senza grande tuo piacere et per tucto di animo, considerando quanta bella forza habbia questa sanctissima unità.

Demostratio xviia a fare tornare allo amico con certe multiplicationi el numero che lui prese di principio

Puossi per forza et virtù di numero fare uno altro mirabile giocho apresso a lungho perché chi ha intelligentia di mathematica che habbi lecto el nostro Euchlide, padre di philosophi, tali giochi et forza di numero li saranno noto, ma omni uno poi non intende ei secreti sui, pure el gioco è sì bello et dilectevole, comme a dire se uno tuo amico pigliassi uno numero di che spetie si voglia, o rotto o intero, quomodocumque sit. Et tu dirai: "io te lo farò multiplicare et

1 non ti bisogna] F non bisogna 14 gettato] F gettando 19 che 'l mia] F del mio 22 che hai] F che tu hai 23–24 dico che non] F dico non 33 sui] F suo

partire per modo che io ti farò tornare | alle mani el numero che tu pigliasti". 29v
Farai così: dirai allo amico tuo che quello suo numero multiplichi per 3 et poi el producto parta in dua et poi di novo multiplichi quello ne viene per 3 et poi el producto parta per 9 et quello ne viene multiplichi per 2 et questa multiplicatione sarà sempre el numero che l'amico prese di principio. Et mai questa regola ti fallirà. Comme se lo amico havessi preso 10, multiplicato per 3 farà 30 et questo 30 partirai per 2, ne verrà 15 et questo 15 multiplicalo per 3, farà 45 et questo dico parta per 9, ne verrà 5. Hora multiplica questo 5 per 2, farà 10, per lo numero prese l'amico tuo di principio. Et se bene l'amico harà preso 2/3, multiplicato per 3 farà 2 et partito per 2 ne virrà 1, multiplicato per 3 farà 3, partito 3 per 9 ne verrà 1/3, multiplicato 1/3 per 2 farà 2/3. Eccoti che la regola serve quando bene fussi rotto et intero insiemi. Comme se l'amico pigliassi 4 1/2 multiplicato per 3 farà 13 1/2, partito per 2 ne viene 6 et 3/4, multiplicato per 3 farà 20 1/4 et partito per 9 ne virrà 2 1/4, multiplicato per 2 farà 4 1/2, per el numero prese. Che in verità sarà giocho | mirabile apresso di ciaschuno che 30r
harà in sé qualche spirito di gentilezza. Et noti ciaschuno che questo libro è facto, scripto et composto per coloro a chi piace, ché so che egli è al mondo di omni sorte de homini: chi si dilecta de una cosa et chi di una altra, ché qualche uno sarà che di simile materia non si delecterà. Et perhò non è facto per quelli a chi non piace et chi di simile materia se dilecterà so che questo mio libretto li fia molto grato. Et se andrai innanze leggendo, troverai ancora cose più utili che queste et di più proficto et di più consolatione, ché el magior piacere che sia in questa scientia è el piacere dell'animo tuo. Che quando tu considerrai poter per forza et virtù di numeri fare si admirande cose et che tu vieni alla opera per te medesimo, certamente è grandissima consolatione di animo et di mente. Comme con l'occhio misurare una torre, una distantia di uno loco, una largheza di uno fiume et molti altri et varii mirabili casi potrai, leggiendo, vedere, siché non ti paia faticha andare leggendo insino al fine et poi piglia a mettere in opera quello che più ti piace et dilecta. Andiamo innanze.

Demostratio XVIIIa a ffare venire alle mani allo amico che numero si voglia 30v
et pigli che numero vole

Io ti voglio dare uno legiadro modo di fare una bellissima et mirabile prova di uno numero per el tuo amico preso et pensato, la quale sia questa. Dirai così allo amico o parente: "piglia qual numero vuoi che ti farò venire alle mani 100

1 che io ti farò] F te lo farò 2 farai] F et farai 7 verrà] F viene 10 farà] F fa 14 farà] F ne verrà 17 piace] K e piace 20 non piace] K e non piace 27 et varii mirabili] F mirabili et varii 28 al] F alla 33 el tuo amico] K e el tuo amico

apuncto medianti certe mia multiplicationi che sarà gioco bello nel conspecto di tucte le persone". Et fassi a questo modo: pigli l'amico uno numero a suo modo o intero o rotto non inporta o rotto solo non ti dà noia. Allora li dirai che multiplichi questo suo numero per 5 et poi quello fa multiplicare per 4. Et poi dirai che questa somma parta per el numero prese di principio et quello che ne viene, dirali, lo multiplichi per 5. Et sempre farà 100 apuncto, ma se lo amico havessi preso 10, multiplicato per 5 fa 50 et poi questo 50 multiplichi per 4, farà 200. Hora dico che questo 200 lo parta per el numero prese di principio che in questa lo harai a partire per 10, verranne 20. Hora dico che questo 20 lo facci multiplicare per 5 farà apuncto 100. Eccoti che la regola riesce et 31r ha forza mirabile. | Hora se lo amico piglassi numero con rotto, medesimamente ti servirà. Comme se lo amico piglassi 4 1/4, multiplicato per 4 o per 5 non inporta quale numero di questi dua tu multiplici prima. Ora diciamo lo multiplichi prima per 4, farà 17 et questo 17 lo multiplichi per 5 farà 85 et questo parti per el numero prese di principio per 4 et 1/4, verranne 20. Et questo multiplicato per 5 farà 100 apuncto. Et quando bene pigliassi uno numero rotto solo, comme a pigliare 1/2 multiplicato per 4, farà 2 et questo 2 multiplicato per 5, farà 10. Hora dico che questo 10 lo parta per el numero prese, cioè per 1/2, ne verrà 20 et questo multiplicato per 5, farà 100 apuncto. Hora ti voglio dare el fondamento di questa regola donde nasce perché tu lectore possa fare questo giocho in più varii modi. Io te ho decto nel caso prefato che tu facci multiplicare el suo numero in 4 et poi in 5 che lo fai multiplicare in 20, ché tanto è a dire: "multiplica una unità per 4 et poi per 5" quanto a dire: "multiplichala per 20", perché 4 via 5 fa 20. Dunque 4 et 5 sono el ripiegho di tucti. Comme multiplica 3 per 4 farà 12, et poi lo multiplica 5 via 12 farà 60, et poi multiplica 3 via 20 farà 31v 60 et tanto è. Et così an|chora "multiplica uno numero per 3 et per 4" è comme a dire: "multiplicalo per 12" et così: "multiplicalo per 4 et per 6" è comme: "multiplicallo per 24". Et così a multiplicare uno numero per 3 o per 8 è comme dire: "multiplicalo per 24" perché 3 via 8 fa 24 et così a moltiplicallo per 2 et per 12 è quello medesimo perché si multiplica nel suo ripiegho. Et questo medesimo averrà nel partire che partire uno numero per 4 et per 6 è a dire: "partilo per 24". Et così in tucti i numeri che hanno ripiegho ché chi parte uno numero per tre et per 6 è comme partirlo per 18. Dunque nel sopradecto caso tu fai multiplicare el numero dello amico per 4 et per 5 che vol dire tu lo fai multiplicare per 20. Dunque per omni unità che piglierà nel suo numero, per questa tua multiplicatione li farai venire alle mani 20. Et se lui harà preso 4 unità, li verrà 4 volte alle mani 20 et se 6 li verrà 6 volte 20 alle mani. Dunque dicendoli poi che questa multiplicatione, et di 4 et di 5, la parta per el numero prese di principio, per

---

3 non inporta] K non porta

forza li tornerà sempre alle mani 20. Potrai tu da te mutare modo. Comme tu fai di 20, poi tu hai a fare di 24, di 30 che è. El ripiegho di 24, comme è dicto, | è 4 32r
via 6 et 3 via 8 et due via 12. Et el ripiegho di 30 è 5 via 6 et 3 via 10 et 2 via 15, etc. Et quando l'amico tuo harà preso uno numero, vorrà, verbi gratia, che tu li facci venire alle mani 20 et tu voglia operare el ripiegho di 24, terrai questa regola generale: sempre parti el numero che lo amico dice voleva havere alle mani per el numero dello ripiegho che tu operi. Comme in questa: tu vuoi a operare el ripiegho di 24 et voli farli venire alle mani 20, di che tu parta el numero che tu li vuoi fare venire alle mani, cioè 20, el numero del tuo ripiego, cioè per 24, che hai a partire 20 per 24, verranne 5/6. Et questo è sempre el tuo capo che ti farà venire el quesito dello amico. Hora diciamo che lo amico habbi preso 5 multiplicato per 3 fa 15 et poi 15 per 8, 120. Et così se tu lo multiplicassi per 6 via 5 farà 30 et poi 4 via 30 farà 120. Et tucto torna in uno. Hora dico che lo facci partire per il numero prese di principio, cioè per 5, verranne 24 et mai ti fallirebbe. Piglia che numero tu voi ché è infallibile. Hora questo 24 dirai lo multiplichi per 5/6, farà | apuncto 20 per el numero dallo amico chiesto et dimandato. Et così 32v
ti riuscirà quando tu usassi el ripiegho di 30, partendo la somma per il numero prese, li verrà sempre 30 alle mani. Et così, usando el ripiegho di 3 via 9, cioè di 27. Et cosi infiniti da te medesimo ne potrai porre et cerchare. Et in su questa regola potrai fondare molti amirandi giochi che non mi voglio distendere ché saria troppo lungho. Et anche so che el mio Magnifico Juliano ha in sé sì gentile spirito che li basta uno minimo cenno a comphrendere cose più alte et più sublimi che non sono questi nostri ludi, ne se admiri alcuno. Se con tanta fiducia dico: "el mio Magnifico Juliano", perché essendo la sua antiquissima et nobile progenie suta sempre patrona della mia antecessori, così voglio et intendo sia patrone di me minimo in fra li altri. Et sono certissimo che sua Illustrissima Magnificentia non si sdegnerà acceptare sì piccolo dono perché, comme disse nel sacro evangelio el nostro sommo Redemptore: *qui se humiliat exaltabitur*— quanto più humile sirà el dono | tanto più sirà *exaltata sua ampla fama et gloria* 33r
quam Deus ad vota feliciter.

Demostratio XVIIII a ffare senpre tornare allo amicho uno numero da te datoli et faccia lui del suo numero quello vole

Uno altro modo bello di fare uno gentile giocho per forza di multiplicatione et partimenti. Comme dirai a uno amicho che pigli uno numero comme a llui pare, che non ti curi o rotto o intero, a suo piacere. Et poi dirai che questo suo numero lo multiplici a suo modo et partalo a suo modo che lo farai sempre partire per tal numero che li verrà alle mani uno numero da te specificato,

2 ripiegho] K ripieno 22–23 sublimi] K submile

cioè che tu li dirai et ti verrà alle mani tanto. Farai in questo modo: dirai allo amico che pigli uno numero a suo modo, non fa caso, et che di quello numero ne facci quello lui vole et multiplichilo et partalo a suo modo. Sempre tu li dirai che partendolo per quello che prese di principio, quanto li verrà alle mani. Et dirai sempre ti dica per quanto lo multiplica o parte. Et tu sempre multiplica e' numeri che lui opera multiplicare et partire l'uno nell'altro. Et quando lui ferma et tu ti ferma et dirali che partendo el numero ultimo che lui ha alle mani per
33v il numero che lui prese di principio che li | verrà alle mani la tua somma che hai alle mani tu. Eccoti lo exemplo: se lo amico harà preso 5 et che lo amico tuo prima lo multiplici per 6, farà 30 et di poi lo multiplici questo 30 per 5 farà 150 et poi questo 150 multiplici per 4 farà 600, et poi questo 600 multiplichi per 2 farà 1200. Hora dica lo amico tuo: "io non voglio andare piu innanzi". Hora a te bisogna sapere tucti e' numeri che lui ha multiplicato el suo numero. Et prima tu sai che lui ha multiplicato prima per 6 et per 5. Dunque multiplica 5 via 6, farà 30. Hora questo 30 multiplichalo per l'altro numero che lui multiplicò, cioè per 4, farà 120 et questo 120 multiplicarai per l'altro numero che lui multiplicò, cioè per 2, farà 240. Hora tu sai che non hai più numeri. Hora fermati qui et dirai allo amico tuo che parta el suo ultimo producto per il numero che lui prese di principio, cioè per 5, ne verrà 240 che in questa hai a partire 1200 per 5, ne verrà apuncto 240. Et mai ti fallirà questa regola. Ancora potresti dire che partisse la sua ultima somma per el numero che tu hai alle mani et che li verrà el numero
34r che lui prese di principio comme in questa. | Se lui partissi 1200 che lui ha alle mani per 240, li verebbe apuncto 5. Et quando bene lui pigliassi numero rotto comme 6 1/5 multiplicato per 3, prima farà 18 3/5 et poi per 5 farà 93 et poi per 10 farà 930, et questo per 2 farà 1860. Et tu da te multiplicha le figure che lui ha multiplicato el suo numero che prima hai a pigliare 3, ma non porta che tu ti cominci così da 3 che tanto farà a cominciarsi dal primo comme dal'ultimo, tanto farà, siché comincia te dove tu vuoi: puoi cominciare dal 2 et dire 2, et poi 10, dirà 2 via 10 fa 20 et questo multiplicha per 5 farà 5 via 20, 100, et questo 100 multiplicha per l'altro numero, cioè 3, farà 300 apuncto. Hora dico che questo è el tuo ultimo producto, dirai che parta el suo ultimo producto per quello che lui prese, cioè parta 1860 per 6 1/5, ne verrà 300 apuncto per el numero tuo. Et così li dirai che se lui partirà el suo ultimo producto per 300 che tu hai alle mani tu che ne verrà a llui alle mani el numero che lui prese. Et mai ti fallirà questa bellissima regola operandola bene et potrai fare con epsa aurea regola vari et
34v innumerabili giuochi di gr|andissimo piacere apresso la brighata.

8–9 che hai alle mani tu] K che tu hai alle mani tu 14 prima per 6] K prima 6 24 6 1/5] K, F 6 1/3 32 6 1/5] K, F 6 1/3 35 con epsa] F con questa

Demostratio xxma a fare di uno numero quella operatione vole sempre saperà quanto lui habbi alle mani per certa legiadra et bella forma

Una forza mirabile veramente intendo in questa nuova dimostratione declararti. Et benché in sé habbi qualche parte di difficultà, pure me sforzerò dare più aperta notitia di sua forza et virtù che a me fia possibile. Et nota el dicto del tuo peripatetico Aristotele che haveva sempre in boccha questa parola: *circha difficilia versatus virtus*, le cose che sono note da ciaschuno facilmente si possono conphrendere, ma le cose ardue et difficili non senza ansietà di animo et ingenio si ha la loro vera notitia. Et quanto una cosa è più obscura et confusa, tanto più laude merita quello che si affatigha di havere di quella vero lume. Et pertanto el nostro Aristotele, padre delli peripatetici non diceva tali parolle senza causa. Ma Alexandro Magno soleva dire: *nihil* | *difficile volenti*— 35r
che quando uno vole, nulla li è difficile. Ma bisogna operare quella via et modi per li quali tu possa venire alla tua desiderata voglia. Ma qui è chi dice: Alexandro Magno quale era Signore dell'universo poteva dire così ché a llui non era difficile cosa alcuna. Comme anchora può questo medesimo vostra Illustrissima Magnificentia, alla quale nulla cosa è difficile poterla conseguire. Ma noi, humili plebei, non possiamo dire così ché qualche volta desideramo cose inpossibili a poterle fare et mai obtenere. Ma li homini virtuosi non desiderano se non cose che mediante epsa virtù potranno facilmente venire alle loro desiderate voglie. Hora attendiamo alla nostra materia. Diremo così: dirai a uno tuo amico che pigli in quore uno numero di che sorte et qualità si voglia, ma nota che in questa regola non bisogna vi sia rotti, che sia numero intero. Hora dirai che di questo suo numero ne facci che operatione a llui pare, a te incognita, che non t'inporta. Cioè | che multiplichi per qual numero si vole et che lo parta 35v
per quale numero lui vole et pongavi et levi quella quantità che li pare et che non ti dicha niente ché non ti curi sapere sua multiplichatione o partimenti. Et questo facci tante volte quante a llui è di piacere, ché se bene lui durassi uno di intero, non ti darà noia. Hora quando lui è fermo et che non vorrà più oltra procedere et tu dirai così che vuoi che una volta o dua al più, lo multiplichi a tuo modo questo suo ultimo producto. Et nota che a te bisogna fermarti in su uno numero che sia tuo partitore et multiplichatore et che questo sia numero terminato in zero, cioè in 0 comme 10, 20, 30, 100 et 1000. Ma per più facilità tua et dello amicho, piglia 10 et quando lo amicho è fermo che dirà: "io non voglio più procedere col mio numero", hora dirai: "io voglio che tu multiplichi questo tuo ultimo producto per 10". Et nota che per questa sua multiplichatione del suo producto in 10, li tornerà alle mani uno numero terminante in nulla, cioè

3 una forza mirabile veramente] F una forza veramente mirabile   33 cioè] K ciò, cioè

36r in zero, perché omni unità in sé | multiplicata per 10 produce 10. Et per forza omni numero che lui havessi alle mani tornerà in figura del'ultima di 10, cioè in 0. Et tu per coprire allo amico la tua multiplicatione di 10 dirai lo multiplichi nel ripiego di 10, cioè in 2 via 5, fa 10. Et comme altra volta ti ho dicto, tanto è a multiplicare uno numero per 2 et per 5, quanto a multiplicarlo per 10. Hora comme lui ha facta questa multiplicatione per 10, tu sai che lo amico ha per sua ultima figura uno zero, dirali vi ponga su uno numero a suo o tuo modo, ma più presto a tuo ché più acomodamente potrai venire al tuo intento. Ma a ciò meglio possa intendere daremo lo exemplo. Poni che l'amico habbi preso di principio 7 et che questo da sé medesimo lo multiplichi per 4, fa 28 et questo 28 multiplicato per 6 farà 168, et questo multiplicato per 3 farà 504, et questo 504 partito per 4 verranne 126 et questo parti per 6, verranne 21 et questo 21 poi multiplichi per 3, farà 63, et decto 63 multiplicato per 2 farà 126. Hora diremo lui non voglia più procedere et fermise qui et dica: "io ho uno mio ultimo pro-
36v ducto et non voglio | più ire avanti". Allora tu dirai che el suo ultimo producto lo multiplichi per 10, overo per 2 et per 5—suo ripiegho, ché tanto farà dove multiplicato 126 per 10 o suo ripiegho, farà 1260. Hora per le tue multiplicationi tu sei certo che l'ultima figura del suo numero è zero, perché l'hai facta multiplicare per 10. Adesso dico che li facci porre in su la sua ultima somma uno numero a tuo modo che a te sia noto et pognamo sia 9. Et tanto dirai ponga in su la sua somma et sai che lui ha 9 per l'ultima sua figura che l'amico harà alle mani 1269. Hora dilli che multiplichi questo suo numero per quanto vole una altra volta, dandoli la libertà, non ti privando, perhò, di poterlo fare multiplicare una volta a tuo modo. Ma questa seconda volta bisogna ti dica per quanto lo multiplicha et per quanto lo parte et tu vi tieni el capo a bottegha. Et nota tucti li ultimi producti che può fare el tuo 9. Che diciamo lui multiplichi questa sua ultima somma per 4, cioè 1269 via 4 farà 5076 et sai certo che 9 multiplicato
37r per 4 produrrà per sua ultima | figura 6, perché 4 via 9 fa 36. Dunque tu sai per certo che l'ultima sua figura dello amico è 6. Ora dirai se più li vole multiplicare. Che poniamo dica: "io lo voglio multiplicare per 2" che multiplicato 5076 per 2 farà 10152. Hora per questa seconda multiplicatione tu sai che l'ultima figura dello tuo amico è 2 perché 6 multiplicato per 2 produrrà per sua ultima figura 2, che 2 via 6 fa 12. Hora lo amico dirà non volere più andare innanti ché se vol fermare qui. Che quando volessi andare più avanti tu, hai sempre a tenere l'occhio al'ultimo producto della sua ultima figura ché se multiplicassi questa sua ultima somma, verbi gratia, per 7 tu sai che la sua ultima figura è 2, multiplicata per 7 diventerebbe 4, et così in infinito. Hora quando lui è fermo, per venire

---

9 poni] K poni poni 27 farà] F fa 28 4 via 9 fa 36] F 4 via 6 fa 36 31 farà] F fa 35 che se multiplicassi] F che multiplicassi

allo effecto del giucho, dirai che questo suo ultimo numero che lui si trova, lo parta per 10 che partito 10152 per 10 ne verrà 1015 et avanzerà 2. Hora dirai allo amico che lassi andare quello ne viene et pigli quello li avanza. Ecco che tu sai certo che lo | amico tuo ha alle mani 2 per tua sottile investigatione. Hora dirali 37v
che questo suo avanza li resta, che tu ne se' certo che lui ha 2 alle mani, che lo multiplichi per quello numero a llui pare et che lo multiplichi per quante volte lui vole, purché a te sieno note tucte sua operationi et che lo parta a suo modo. Ma a ciò lui non comphrenda che ne habbia tale notitia, potrai fare così: cioè farlo multiplicare una volta o 2 a tuo modo, tanto che lui non possa comphrendere quello che tu voglia fare, et poi potrali dare la libertà a suo modo che lo multiplichi et parta et giungha et scemi comme a llui pare. Et tu sempre li sarai alle falde et sempre saprai quanta somma lui harà alle mani. Et comme lui si ferma che più non voglia procedere et tu li dirai: "tu hai alle mani tanto", verbi gratia noi diciamo che li avanzò 2. Hora diciamo che lo multiplichi per 4 et farà 8 et questo 8 multiplichi per 6 farà 48 et questo parta per 3, verranne 16 et questo multiplichi per 5, farà 80 et questo multiplichi per 10 farà 800 et questo parta per 8 verranne 100. Hora, se tu harai tenuto l'occhio | alla pentola, tu saprai che 38r
lui ha alle mani 100, il che non ti potrà l'amico tuo negare ché in verità è gioco mirabile nel conspecto di ciaschuno. Et nota che in questo giocho ci va più forza et virtù de ingegno che di numeri perché poco ci adopra la forza di numeri in questo caso. Ma quando tu hai dato el numero allo amico, cioè quando tu lo hai facto multiplicare per 10 et che tu li dai el numero tuo comme tu li desti 9, se lui multiplica poi el suo ultimo producto per 10 overo per numero terminante in zero, ecco che tu non potresti venire al tuo disegno ché non potresti sapere el suo ultimo producto. Et perhò ti darò modo a questo. Quando lui lo vole multiplicare per numero terminante in zero comme per 10, 20 o 30 etc., dico che in questo caso tu li dica non volere che lui proceda più oltra con la sua seconda multiplicatione perché tu hai a notare che in questo gioco c'è tre multiplicationi: la prima che lui fa tacite da sé preso che lui ha el numero; la seconda quando tu li hai dato el numero che di questa bisogna che tu ne habbi notitia; la tertia quando tu li hai facto partire | per 10 che tu vieni alla notitia di 38v
quanto lui si trova et che tu li dai la libertà a briglia sciolta di multiplicare a suo modo. Dunque, tornando a proposito, nella seconda multiplicatione che lui fa, dandoli la libertà ché fia tenuto più admirando se tu li dai la libertà, et lui lo multiplichi per numero terminato in nulla, dirali si fermi che non proceda più oltra ché procedendo poi sopra la prima multiplicatione non potresti investigare el bisogno. Et perhò farai si fermi et comme è fermo, terrai questo modo: a te bisogna notare quando lui multiplica, per che numero lui multiplica, se lui

9 a tuo] K a tuo a tuo 33 dunque] F adunque

multiplica per 10 o suo ripiego. A ciò meglio intenda, pognamo l'amico habbi alle mani 7 per sua tacita multiplicatione, cioè per sua prima multiplicatione, dove se tu lo farai multiplicare per 10, harai 70. Et poi li darai uno numero a tuo modo, diciamo li dia 9, harà alle mani 79. Hora diciamo lui multiplichi questo 79 per 10 farà 790. Adesso tu sai che el tuo 9 multiplicato per 10 produce 90 et perhò dirai parta el suo numero che lui ha per 100, verranne 7 et resterà 90.
39[r] Dirai che la|ssi ire el 7 et pigli 90. Et farrai quelli medesimi efecti decti di sopra et se lui multiplichassi per 20 o per 30, comme per 30 el tuo 9 produrrà per le 2 ultime figure 70 che 9 via 30 farà 270, facendolo partire per 100 sempre li avanzerà 70, perché 30 via 79 farà 2370, partito per 100, ne verrà 23 che non si cura et avanzerà 70. Et così farai per tucte le figure che lui multiplichassi. Tu sai che la sua ultima figura è zero et perhò noterai la penultima. Et quando lui multiplichassi el suo numero per 100 o suo ripiegho, cioè per figura havencte 2 zeri et tu farai poi partire el suo producto per uno numero havente uno zero più che il suo multiplicare. Comme se lui multiplica per 10 et tu fai partire per 100 et se lui multiplica per 100 et tu fai partire per 1000, et se lui multiplica per 1000 et tu fai partire per 10000. Et sempre nota la figura che viene a llato al zero. Comme se lui multiplicarà 79 che lui ha alle mani per 100 farà 7900, dove dirai che parta 7900 per 1000, verranne 7 che non si cura et avanza 900, dove tu sai che el tuo
39[v] 9 multiplicato per 100 produsse 900. Ecco | che tu harai la medesima notitia. Multiplichi per quanto lui vole et se multiplicarà per 200 o per 300, hai a notare che numero produca el 2 o el 3, ché se lui multiplicarà 29 per 200 tu sai ch'el 2 multiplicando col tuo 9 produrrà per sua ultima figura 8 et perché multiplicato per 100 tu farai partire per 1000, ne verrà 15 che non si nota et avanzerà 800, il che sapiendo, tu potrai fare li tui effecti. Potrai in su questa dimostratione et regola fondare molti et varii giochi di tua fantasia che saria troppo lungho a volerle tucte in questo volume dire, ma tu col tuo mirabile ingegno potrai tucti conphrendere et questo basti quanto a questa demostratione.

Demostratio xxia della forza del multiplicare

Non voglio preterire in questa nostra ultima demostratione di questa prima parte, di mostrare la forza et virtù del numero, quando la sia anchora nel multiplicare e' numeri insiemi uno coll'altro, ché in verità fassi cose mirabili, et sono
40[r] cose da tenerle quotidianamente alla memoria che fanno | honore, mirando nel conspecto universale delli homini. Comme: trovami 2 numeri che multiplicati uno nell'altro mi produca 6 figure che sieno tucte in uno modo, comme tucti 1, o tucti 2, o tucti 3. Ma nota che queste figure non vogliono passare 6 figure, perché queste sei figure con una regola breve, che io ti darò, potrai fare sopra

---

27 dire] K dirle 30 preterire] F pretretirire 37 sopra] K lopra

giornalmente et averla in memoria. Quando volessi passare el numero di 6 figure, daremo modo di sotto. Ma nota queste 2 regole prima ché sono belle et facile. Darotti dua modi facili de' quali el primo fia questo: tu hai a tenere per tua guida in questa prima regola 2 numeri: el primo è 777 che questo è immobile ché sta sempre firmo; l'altro è 143 et questo è mobile secondo el bisogno che vuoi fare. Dico adunque che se tu vorrai fare 6 figure tucte unite, che tu multiplichi 777 via 143, farà 111111. Ecco che la regola riescie a ffare le unità. Hora, quando tu volessi fare tucti 2, cioè 6 "2" comme 222222, in questo dico che tu adoppi 143, farà 286, et questo | multiplicha via 777 immobile, farà el numero [40v] quesito. Et se tu vorrai tucti 3, piglierai 143 3 volte, che farà 429, che multiplicato 429 via 777 farà 333333. Et volendo fare venire tucti 4, piglierai 4 volte 143, farà 572, che multiplicato via 777 farà 444444. Et se tu volessi tucti 5, piglerai 5 volte 143 che farà 715, et questo multiplica via 777, farà 555555. Et se tu vorrai tucti 6, pigliarai 6 volte 143, farà 858, qual multiplicato via 777 farà 666666. Et se tu volessi tucti 7, piglerai 7 volte 143, farà 1001. Et questo multiplica via 777, farà 777777. Et se tu vorrai tucti 8, piglerai 8 volte 143 farà 1144, qual multiplicato per 777 farà 888888. Et se tu vorrai tucti 9, piglerai 9 volte 243, farà 1287, qual multiplicato per 777, farà 999999. Et così hai. In tucti e' modi può venire.

Hora nota l'altra bella et perfecta regola. Potrai anchora fare a questo modo: tu vorrai fare venire tucte unità, farai così: del numero che tu vuoi fare venire, guarda la figura che tu | vuoi fare venire quanto la dimostra et pigla per omni [41r] unità che la dimostra 2 centinaia et poi tre decine per omni unità similiter. Et tucto acoza insiemi. Et a questo agiugni la figura che tu vuoi che vengha et poi multiplica tucto questo per uno numero immobile che fia questo: 481. Et sempre harai el tuo numero certo. Et è regola generale quando tu voi fare venire tucte de una ragione, cioè o tucte unità, o tucti 2, o tucti 3, etc. ... Comme se tu vorrai fare venire tucta unità farai così: guarda la unità representa uno, dunque hai a pigliare 2 centinaia. Hai 200 et hora 3 dicine, sono 30, poste sopra 200 fa 230. Hora agiugni la figura quesita, cioè 1, farà 231. Et questo multiplica via 481, farà el numero quesito, cioè 111111. Et se tu volessi fare tucti 2, dico che guardi la figura che tu vuoi fare quante unità rapresenta, che rapresenta 2. Dunque hai a pigliare 2 volte 200, fa 400. A questo poni per omni unità che rapresenta la figura 3 dicine et in questa hai a pigliare 6 dicine, | sono 60, agiuncte a 400 [41v] fa 460 et a questo agiugni 2 che fa 462. Et questo dico che multiplichi via 481, farà 222222. Et se tu vorrai tucti 3, piglerai per omni unità 200 che hai 600, et poi per omni unità 30 che hai 90, giugni insiemi fanno 690, postovi su 3 farà 693, multiplicato via 481, farà 333333. Et quando vorrai tucti 4, tucti 5, tucti

6 tucte unite] K tucte unita 25 el tuo numero certo] K el tuo numero cercho

6, sempre per omni unità che rapresenta la figura che tu vuoi, hai a pigliare 200. Et poi per omni unità similmente 30, et poi el numero che tu vuoi che venga et tucto acozerai insiemi et multiplicarai via 481 et harai el tuo numero. Comme se tu vorrai tucti 7, hai a pigliare 7 volte 200, fa 1400, et poi 7 volte 30 fanno 210, posto sopra 1400 farà 1610, et poi 7 farà 1617 et questo multiplica via 481, farà 777777. Eccoti dua modi brevi da tenerli quotidianamente alla memoria. Ma se tu volessi fare venire numeri tramezati, comme fare venire 3 et 2 in questa forma: 323232, overo 4 et 6 comme: 464646, overo 5 et 9 comme: 595959 o
42r numeri simili a questo, non passando co|mme è dicto di sopra 6 figure in tucta la somma della multiplicatione, et quando volessi passare 6 figure, ti darò di sotto breve modo, ma alquanto più difficile che li precedenti, perché dove va quantità di figure, non si può fare che in sé non habbino qualche parte di difficultà. Pure el bono abachista et pratico a multiplicare et partire non li sarà nulla difficile. Terrai dunque questo modo: quando tu vorrai figure tramezate comme 3 et 4, overo 4 et 5, overo 7 et 9 o simili, dico che ti bisogna notare quante unità rappresentano le due figure che tu vuoi fare venire et per omni unità che le rappresentano per regola generale sempre pigla 20 et somma insiemi. Et poi vi poni su le due figure che tu vuoi che venghino. Et questo pure dico che multiplichi via 481 numero immobile comme nella regola superiore. Hora diciamo che tu voglia fare venire 252525, dico che tu guardi che 25 rapresenta 25 unità. Dunque piglierai 25 ventine che saranno a numero 500 sopra questo, bisogna che tu
42v ponga 25, cioè le due figure che tu vuoi | che venghino, farà 525. Et questo ti bisogna multiplicare via 481, farà 252525. Et se tu vorrai 343434, piglierai 34 volte 20, che farà 680, et sopra vi poni 34, farà 714. Et questo multiplicharai via 481, farà 343434. Et se tu volessi 505050, piglerai 50 ventine, seranno 1000 apuncto, et a questo agiugni 50, cioè le 2 figure, farà 1050 et questo multiplicharai via 481, farà 505050. Ecco che tu hai el modo facile a tenerlo quotidianamente alla memoria che sempre ti riuscirà, tenendo queste date regole che farai venire qual generatione di figure che tu vorrai, o 707070 o 191919, in qualunché modo ti piacerà, non passando 6 figure comme è dicto. Ma quando vorrai fare venire più che 6 figure terrai questo modo: quando tu volessi trovare 2 numeri che multiplichino l'uno nell'altro te facessi venire 12 unità, overo 12 dua, overo 12 tre, comme a questo modo 111111111111, overo a questo modo 222222222222,
43r overo a questo modo, cioè 444444444444 o | simili numeri, farai tuo fundamento in su questo numero, cioè 900991. Et questo 900991 sarà sempre tua guida in tucti i numeri che tu volessi fare venire, cioè partirai quello numero che tu vuoi che vengha per 900991 et quello che ne viene sarà tuo multiplica-

2 similmente] F *omissis* 12 quantità] F quanti | habbino] F habbi 30 venire] K avenire 32 ti facessi venire] F ti facessino venire

tore, cioè multiplicharai quello che viene dicto partimento via 900991 et farà el numero comme vuoi che diciamo tu voglia 12 unità, partirai quelle 12 unità per 900991, verranne 123321 et questo multiplicarai via 90091 farà 111111111111. Et se tu volessi 12 dua, partirai quelli 12 dua per 900991, verranne 246642, et questo multiplicharai via 900991 et farà 222222222222. Et così, quando vorrai 333333333333, farai nel modo dato. Et anchora se tu volessi numeri tramezati comme 232323232323, ovvero 828282828282 et simili, sempre parti el numero che tu vuoi che ti vengha per questo 900991 et verratti el numero quesito. Et questa è regola generale quando tu volessi 100 unità o 100 dua etc. Ma questo multiplicare o partire per così grandi numeri me pare | materia fastidiosa, perhò 43v non procederò più avanti con questa mia demostratione. Et fanno fine a questa nostra prima parte, dove havemo dati li sopradecti modi a trovare uno numero intero non diviso in parte alcuna. Hora ti voglio dare modo a potere investigare uno numero diviso in più parte che in verità non saranno mancho sottile delle sopradecte né di mancho piacere, pure saranno le cose nostre tenute di piacere da quelli homini che di simili arte si dilecteranno ché tucte le arti dilectano a tucti li homini, ma omni arte non dilecta omni uno et che se tu dimandarai me se la scientia mathematica mi dilecta, fia comme colui che dimanda quello sarto se lui sapeva fare uno mantello o uno barbiere se sa cavare denti, siché non bisogna domandare omni uno sempre la scientia che lui si dilecta. Quella sempre exalta et magnifica, ché sempre el philosofo dirà la philosofia essere la prima scientia del mondo, lo astrologho dirai l'astrologia, el musico la musicha, l'arismetrico l'aris|metricha et si è de singhulis. Ma el contrario nasce nelle arte 44r mechaniche ché se tu domandi uno lanaiolo se la sua arte è bona, dirà che è meglio fare el votacessio. Così se tu dimandi el setaiolo, dirà non è al mondo la più poltrona arte et in effecto dice uno proverbio vulgare: omni uno ha bona moglie et captiva arte. Questo se intende a decto di lui, onde dice el tuo Oratio nel principio della sua satira, scrivendo a Mecenate, dice così: *Quid fit Mecenas ut nemo quam sibi sortem, seu ratio dederit seu sors obicerit, illi contentus vivat et laudet diversa sequentis*? Denota Oratio in questi versi che niuno è contento di sua sorte, o bona o captiva che lui habbi, ma dice Tholomeo: *Sapiens dominabitur astris*, chi ha in sé qualche parte di ragione resta poi contento alla voluntà del nostro sommo factore che omni cosa regge et governa con la sua infinita sapientia. Hora, tornando a nostro proposito chi di questa scientia se dilecta, stimo questa nostra opera | li sia cara. Hora faremo fine a questa nostra 44v prima parte. Anderemo alla seconda.

---

9 unità] F *omissis* 16 da quelli homini] F dagli homini | si dilecteranno] F se dilectano
24 dirà] K dirai 31–32 *dominabitur*] F *dominatur*

Seguita la seconda parte del libro decto Giochi mathematici

Ricordami havere udito Magnifico Juliano dalli mia antecessori et antiqui che intervenne nelli preteriti tempi allo Illustrissimo duca di Milano volere fabricare un ponte in Lonbardia sopra una fiumana decta l'Adda, se bene me ricorda. Et a ciò meglio tale effecto conseguire potessi, convocati tucti li maestri et architectori della sua lonbarda regione, proposta loro tale opera volere mandare ad effecto che desiderava, li dicessino l'ordine et modo che a tale artificio era necessario et che con mancho spendio et più facilità che a lloro fussi possibile li dessino disegno.

Il che inteso dalli prenominati maestri et architectori et sopra ciò preso quella conclusione li parve oportuna, mostrorono alla excelentia del signore duca tal cosa non potersi fare senza difficultà et per questo ne seguitava spese
45r a tale opera inconvenienti. Il che inteso | el Magnifico duca, di tale cosa sbigottito, più et più anni la lasciò inperfecta comme quello che giudicava essere quasi inpossibile.

Advenne che, forte fortuna, capitò nella ciptà di Milano uno architectore fiorentino, il cui nome adesso non l'ho alla memoria. Et, essendo epso nella prefata città, el duca, comme desideroso de intendere et vedere li homini virtuosi, un giorno lo fece ad sé venire et fra li altri electi ragionamenti li dimandò consulta sopra questo ponte che più anni, comme iusto et bono signore et di sui popoli amorevole, haveva desiderato di fabricare, perché a popoli sua viandanti dava grandissima incomodità questo fiume. El nostro architectore fiorentino, inteso et visto con lo spiculativo occhio la cosa di che qualità era, brevemente li respose che con mancho spendio et con molta più facilità tale opera fabricare si poteva che dalli sui architectori li era stato exposto, offerendosi lui in simile exercitio volentieri affatigarsi per fare cosa grata a sua signoria.

El che molto piacque al duca et aceptate le sue offerte dette principio a
45v fabri|care uno magno et spatioso ponte sopra el predicto fiume. Hora, comme adviene, el nome andò per tucta la Lonbardia comme uno maestro fiorentino haveva principiato el prenominato ponte. Onde molti et vari architectori dilectandosi intendere, vennono a vedere el modo e l'ordine che costui haveva trovato. Di che naque che molti o la magior parte, inteso el pregio che costui aveva di dicta opera, andorno al signore duca, dicendo che con mancho priegio assai harebbono loro tale ponte fabbricato che non faceva questo fiorentino. Il che indusse nell'animo del duca un pocho di sdegno che si riputava giunctato et alquanto inanimato contra epso maestro. Et factolo a sé venire, li disse

6 proposta loro] F proposto loro

havere molti maestri et architectori nella sua ciptà che questo ponte harieno per mancho pregio assai che lui conposto et fabricato. Di che lui intendeva che lui venisse a pregi che e' sua architectori erano consueti. Donde el nostro architettore, comme homo experto di acutissimo ingegno, dimandò in gratia al duca che facessi venire quelli homini che di simile cosa | si vantavano [46r] ché voleva con loro conferirse et non era per iscostarsi dalle cose iuste et ragionevole, il che subitamente fu facto. Venuti adunque li prenominati maestri et constituiti nella presentia dello illustrissimo signore duca, insieme col nostro fiorentino, furono i predicti da epso dimandati se erano loro quelli che tale cosa admiranda fare volevano, il che da lloro fu quivi confirmato. Onde el nostro fiorentino, preso in mano uno uovo et postolo in su uno schacchiere pulito, disse alta voce: "horsù, chi di voi ferma questo uovo qui nel mezo di questo schacchiere senza ronperlo et con presteza, cioè innanzi che passi uno quarto di hora, hoggi da me guadagnerà 50 ducati". Onde, tucti stupiti, quivi non fu nessuno che potessi questo effecto mandare ad executione, ma guardando l'uno l'altro si ristrignevono nelle spalle. Onde passato lo spatio da llui dato che passò presto, disse loro: "se la rizassi io nel modo predicto, voi direte bene che io sia valente homo?". Tucti una voce dissono: "sì, certo", perché infacto chi non | ha el modo et sua facilità, li parrà cosa stu- [46v] penda. Onde furno constrecti dire sì. Et lui, preso di uno cantuccio o camino uno pocho di cenere et quella messa in sul mezo dello schacchiere, vi rizzò su l'uovo con la sua puncta. Onde li maestri lombardi cominciarno a ridere, facendosi beffe di simile acto. Dissono che in quale modo anchora loro saprebbono porre l'uovo, di che el nostro fiorentino rispose: "così sapete fare el ponte hora ch'io vi ho mostro el modo, ma prima ch'io vi mostrassi la via, non sapevi fabbricare el ponte et non sapevi rizzare l'uovo nel mezo dello schacchiere. Et omni uno sa fare a giuocho insegnato". Alle quali parolle non fecero risposta alcuna, il che inteso, el magnanimo duca, molto li piacque simile acto et giudichollo homo veramente di mirabilissimo ingenio. Et licenziati li sua architettori, li disse che alla sua opera attendessi. Ho voluto, Magnifico Juliano, questa brieve faccetia recitare, perché forse sarà qualche uno più presunptuoso ch'el bisogno che dirà: "io anchora havrei saputo una simile opera compilare", et diranno questa essere | suta di grandissima facilità, a [47r] che rispondo comme el nostro fiorentino che ciaschuno sa fare a giocho insegnato et che quando la via è facta, omni uno sa per epsa facilmente caminare. Ma io mi confido nel mio protectore et padrone sotto l'ombra del quale so

---

6 conferirse] K conferissi | et non era] K et non era et non era 11 uovo] K ouvo | in su uno] F in su n'uno 12 uovo] K ouvo 14 hoggi da me] K *omissis* 17 se la rizassi io] K se io la rizassi io 22 uovo] K ouvo 24 uovo] K ouvo 26 uovo] K ouvo 35 per epsa] F *omissis*

che non potrò perire et dirò ad voi, Magnifico Juliano, in questa forma: In te Domine speravi, non confundar in eternum et poi: entrerrò in omni profundità senza timore alcuno. Hora attendiamo all'opera nostra et vegnamo alla prima dimostratione di questa seconda parte.

Demostratio prima a trovare uno numero diviso in due parte

Avendo io, Magnifico Juliano, dato nelle superiori dimostrationi via et modo di poter ritrovare uno numero intero o rotto in sé, non diviso in parte alcuna, mi pare conveniente in questa nostra seconda parte dare qualche notitia di trovare uno numero diviso in più parte. Et per potere più ordinatamente procedere, prima darò modo a trovare uno numero diviso in 2 parte et poi in più, seguendo di mano in mano ordinariamente. Sarà adunque el primo modo di trovare uno
47v numero diviso in 2 parte. Comme se uno dicessi: "io ho uno numero | in quore el quale è diviso in 2 parte", ovvero uno dice: "io ho una quantità di danari o grossi o quarteruoli o nociuole o chiose, et pongoli in su nuna tavola et due ne piglia omni uno la sua parte", ovvero omni uno ne tolse quante più potè, overo "io ho in questa mano una quantità" o in omni modo che lla si sia. Terrai questo breve modo, cioè dirai: "pigla uno numero a tuo modo a me noto", qual preso dirà: "io lo divido in 2 parte intere". Et nota che se lui lo dividessi in parte rotte non ti servirebbe questa regola. Et perhò dirai lo divida in 2 qual vol parte intere che non ti dà noia et poi dirali che multiplichi una di dicte parti per 2 et al'altro dirai che multiplichi per tucto lo numero diviso che sia chiuso, diviso 12. Dunque dirai che l'uno di loro adoppi el suo numero et l'altro lo multiplichi per il numero diviso, cioè per 12. Et poi conferischino insiemi loro multiplicationi et acozino tucto insiemi. Poi tu tacite multiplica el numero diviso per uno più di epso, cioè in questa el numero diviso è 12. Dunque multiplica 12 per uno
48r più, cioè per 13 et quello fa. Dirai che cavino el numero loro dal tuo et die|dino quello avanzerà. Allora tu piglierai lo avanzo et partiralo per uno mancho che il numero diviso et quello che ne verrà sirà el numero di quello che adoppiò el suo, et quello che avanzerà sarà el numero di quello che multiplichò per el numero diviso. Hora ecchoti lo exemplo: sia comme è ditto el numero diviso 12, factone due parte, overo che dua fra loro habbino diviso l'una 7, l'altra 5. Hora dirai a uno di loro che adoppi el suo numero che pognamo sia 7, doppiato fa 14. Hora dirai al'altro ch'el suo numero multiplichi per tucto el numero diviso che sarà el 5, multiplichato per 12 farà 60. Et acozato 60 et 14, farà 74. Hora tu da te tacite multiplicha el numero diviso per uno più, harai a multiplichare 12 via 13, farà 156. Et tu dirai loro che cavino la loro somma di 156, che cavato 74 de 156 resterà 82. Et questi dico che parta per uno mancho ch'el numero diviso, cioè per 11, verranne 7 per il numero di quello che radoppiò et avanzerà 5 per il numero di quello che multiplichò el suo numero per il numero diviso. Et così se

quello che haveva 5 havessi doppiato el numero, farebbe 10 et que|llo che ha et havessi multiplichato el suo numero per il numero diviso, cioè per 12, farebbe 7 via 12, 84. Et acozato insiemi 10 et 84, fa 94. Hora tu da te multiplicherai comme di sopra 12 per uno più, farà 156, dove dirai che cavi la sua somma della tua, cioè cavi 94 di 156, resterà 62, quale partito per uno meno che il numero diviso, cioè per 11, verranne 5 per il numero facesti doppiare et avanzerà 7 per il numero che colui multiplichò per il numero diviso, cioè per 12. Eccho che la regola sarà sempre generale, salvo quando una delle 2 parte sia unità. Comme s'el numero diviso fussi 4 diviso in 2 persone, l'uno ne havessi preso uno et l'altro 3. Se adoppiassi la unità farebbe 2 et l'altra multiplichato via 4 farebbe 12, acozato tucto insiemi farà 14, cavato della multiplichatione di 4 in uno più, cioè in 15, cioè di 20, resterà 6. Partito 6 per uno mancho del numero diviso, per 3, ne virrebbe 2, che non è nessuna di dicte parte et saria falsa la regola. Et perhò quando tu non sai certo che una delle parte | non sia unità, dirali che doppi lo magiore et virratti bene, ché in questa se tu doppi 3 farà 6, multiplicata la unità per 4 farà 4 et 6 haverai, prima farà 10 cavato di 20 resterà 10, partito per 3 ne verrà 3 per il numero che facesti doppiare et avanzerà uno per il numero multiplicò per il numero diviso. Et la regola servirà sempre quando le parti saranno intere che non vi sia 1/2 o 1/3, comme potrai experimentare.

48v (margin, line 1) · 49r (margin, line 15)

Demostratio secunda a trovare 3 parte de uno numero diviso

Hora diremo se 12 fussi diviso in 3 parte a te incognite, comme se tu ponessi in tavola 12 quarteruoli et dicessi a tre persone che omni uno ne pigliassi quanti li pare, et tu volessi poi investigare decte parte, terrai questo breve modo. Cioè prima dirai a uno di loro qual vuoi, che non fa caso che adoppi el suo numero. Et al'altro dirai che multiplichi el suo numero per el numero diviso, cioè per 12. Et l'altro dirai che multiplichi el suo numero per uno più ch'el numero diviso, cioè per 13 et tucta questa lor somma dirai che la conferischino insie|mi et acozino et faccino tucta una somma et serbino. Et poi tu da te tacitamente multiplicha el numero diviso comme nella superiore per uno più ch'el numero diviso, cioè per 13. Poi dirai che cavino la somma loro de la tua, et quella resta tu la dividerai per uno meno ch'el numero diviso, cioè per 11. Et quello ne verrà sarà el numero che tu facesti adoppiare, et quello avanzerà sarà el numero di quello che multiplichò per il numero diviso. Et l'altra parte sarà quello mancha insino in 12, sia comme è dicto el numero diviso 12, partito in 3 parte: 3, 5, 4. Hora dirai a uno, che pognamo sia quello che ha 3, che adoppi el suo numero, farà 6. Et poi dirai

49v (margin, line 27)

13 nessuna] F niuna 24 vuoj] F voli 30 de] F et 34–35 a uno che poniamo] F a uno poniamo

a uno altro, sia quello che ha 5, che multiplichi el suo numero per il numero diviso, cioè per 12, farà 60. Et poi dirai a quello altro che ha 4, che multiplichi el suo numero per uno più che il numero diviso, cioè per 13, farà 4 via 13: 52. Hora dirai che tucta questa somma, cioè 6 del primo, et 60 del secondo, et 52 50r del tertio, che tucto acozino insiemi fra loro, farà 118 apuncto. | Hora tu tacite da te multiplicha 12, cioè el numero diviso, per uno più, cioè per 13, farà 156. Et dirai loro che cavino la somma si trovano alle mani, cioè 118. Dicesti loro serbassino et farai ti dichino quello rimane che cavato 118 di 156, resterà 38. Et questo 38 dico che parta per uno mancho che il numero diviso, che in questa lo havrai a partire per 11, dove partito 38 per 11, ne verrà 3 per el numero che tu facesti doppiare et avanza 5 per il numero di quello che multiplichò per il numero diviso per il terzo, vedi quanto resta a cavare 3 et 5 di 12 che vi resterà 4. Dunque dirai che il primo ha 3, l'altro ha 5, l'altro 4. Et se tu havessi decto a quello che ha 5 che doppiassi el suo numero, fa 10. Et poi a quello che ha 4 che multiplichi el numero per lo numero diviso, cioè per 12, fa 48. Et poi havessi decto a chi ne ha 3 che multiplichi el suo per uno più ch'el numero diviso, cioè per 13, fa 3 via 13, 39. Acozato tucto insiemi 10, 48, 39, fa 97 et cavato 97 dalla tua 50v somma, cioè di 156, resta 59 che partito 59 per uno meno che il numero | diviso, ne verrà 5 et avanzerà 4, dove 5 che ne viene sarà el numero che tu facesti doppiare et 4 che avanzerà sarà el numero che tu facesti multiplicare per lo numero diviso. Et acozato 4 et 5 fa 9, tracto di 12 resta 3 per la tertia parte et riusciratti in tucti i modi. Et nota che in questa seconda regola non inporta se bene una parte del numero diviso fussi unità et che quella si doppiassi comme nella superiore, comme se il numero diviso fussi 5, partito in 3 parte, cioè l'uno havessi uno, l'altro 1, l'altro 3, dico che facci doppiare l'unità prima farà 2, l'altra unità multiplicala per 5 farà 5. Et l'altro numero, cioè 3, multiplicalo per uno più che el numero diviso per 6, farà 18. Acozato tucta questa somma insiemi faranno 25 apunto, dove dirai che cavino 25 della tua multiplicatione di 5 in uno più, cioè in 6, che farà 30, resterà 5. Questo parti per uno mancho ch'el numero diviso per 4 verranne uno per la prima unità et avanzerà uno per la seconda unità 51r et el resto sino in 5 sarà et l'altro che fia 3. Dareti più exempli, ma | allo acuto ingegno questi dati sono superflui et questi bastino a fondare tucti e giuochi che tu potrai fare. Basta che tu dica che se pigliassino uno numero in quore che non vi sia rotto o parte rotta in nessuna di decte parte, ma sieno tucte unità intere. Et nota che tu potrai fare questo giocho in questa forma. Cioè habbi messo in uno borsellino da canto 10 o 12 grossi o ducati, che non porta o 10 o 30 che sieno, et falli dividere in fra 3 persone levandoti tu di quivi, dicendo omni

---

5–6 tacite da te] F dirai da te 23 diviso] F *omissis* 36 da canto] F dato 37 di quivi] F di qui

uno pigli quanti li pare, cioè sieno 3 a pigliare et poi multplicha comme di sopra et faragli superiori effecti che in verità sarà tenuto bello giuocho nel conspecto di tucti. Hora nota uno altro modo a fare el medesimo effecto.

Demostratio 3a a investigare 3 parte di uno numero diviso

Potrai per una altra legiadra regola et modo venire alla notitia di 3 parte di uno numero diviso per mutare ordine. Et così prima dirai a uno di loro ch'el suo numero lo adoppi. Et poi dirai al'altro che multiplichi el suo numero per uno m|ancho ch'el numero diviso. Al'altro dirai ch'el suo numero multiplichi per il 51v
numero diviso. Et tucte queste multiplicationi le conferischino insiemi in una medesima somma. Et questo serbino. Et poi tu da te tacitamente, per regola generale, quadra el numero diviso, cioè multiplicha el numero diviso in sé et per coprire la regola tua alli circunstanti sopra el quadrato di decto numero, sempre poni uno numero a tuo modo, che non fa caso. Et poi dirai che cavino la somma che a lloro facesti serbare del tuo numero et dichino a te el rimanente. Et poi tu da te cava tacitamente el numero che tu ponesti in sul tuo quadrato, et quello rimane sempre parti per 2 mancho, ch'el numero diviso et quello ne viene sarà el numero che tu facesti doppiare. Et quello che avanza sarà el numero di colui che multiplichò per uno mancho ch'el numero diviso, la tertia parte sarà el restante. Sia el numero diviso 12, diviso in tre parte: uno habbi 3, l'altro 5, l'altro 4. Dirai a uno che radoppi, che diciamo sia quello che ha 4, farà 8. Et poi dirai al'altro | che ha 5 che multiplichi el suo numero per uno 52r
mancho ch'el numero diviso, cioè per 11 et multiplicato 5 via 11 fa 55. Et a quello che ha 3 dirai che multiplichi el suo numero per il numero diviso, cioè per 12, fa 36. Et tucto dirai che acozino insiemi, che fa somma di 99 et serbino. Hora tu multiplicharai da te tacitamente el numero diviso in sé, cioè 12 via 12, fa 144 et sopra questo dico che ponga 10 et quando tu non lo ponessi, riuscirà el giuocho. Questo si fa per ricoprire alli intelligenti el tuo modo. Posto, dunque, 10 in su 144, farà 154. Hora dirai che chavino la lor somma, cioè 99, della tua, 154, resterà 55. Hora tu da te tacitamente cava el numero che ponesti sopra el quadrato di 12, cioè 10, che vi ponesti, resterà 45. Hora dico che parta questo 45 per 2 mancho ch'el numero diviso, cioè per 10, veranne 4, per il numero di quello che radoppiò et avanzerà 5 per il numero di quello che multiplichò per uno mancho del numero pensato per la tertia parte, overo per il tertio numero. Acozerai 4 et 5 insiemi fa 9. El resto per infine in 12 che è 3, sarà l'altro numero. Eccho | che la 52v
regola ti serve. Et se tu havessi decto a colui che ha 3 multiplichassi el suo per 2,

---

12 sopra il quadrato] F sappi el quadrato 16 mancho] K mancho mancho 34 insiemi] F *omissis*

cioè che adoppiassi el suo numero, fa 6 et poi a quello che ha 4, multiplichassi el suo per uno mancho ch'el numero diviso, cioè per 11, fa 4, via 11: 44. Et poi a quello che ha 5 che multiplichassi el suo numero per el numero diviso, cioè 5 via 12, fa 60, et acozato tucto insiemi fa 110. Et questo tracto della multiplicatione del numero diviso in sé, cioè di 144, overo quando tu hai posto qualche numero sopra 144, che questo non è necessario, cioè non bisogna che tu ponga altro numero sopra 144 per fare il giuocho tuo, ma fassi per coprire la facilità sua, tu, lectore, piglierai quel modo, più ti verrà comodo. Per brevità diciamo che cavi 110 di 144, resterà 34. Hora partirai questo 34 per 2 mancho del numero diviso, cioè per 10, verranne 3, per il numero che fu adoppiato et avanzeratti 4 per il numero che fu multiplicato per uno mancho ch'el numero diviso et l'altro numero troverrai nel modo sopradicto agiugnendo le parti sapute, el restante sia l'altra parte.

53r Et non ti admirare, lectore, se io ti ho dato el modo di coprire la facilità di questo giuocho, perché io per me molte volte ritrovandome in qualche loco dove simili giuochi si sono facti, ne ho imparati molti solo a vederlo fare.

Et anchora mi sono abattuto a inparare uno sonetto a mente, a udirlo una sola volta, allegando in testimonio Betto da Cortona, bono, anzi optimo maestro di canto che, saputo che lui hebbe che io inparai uno suo sonetto, non volle mai più cantare in luogho dove io mi trovassi, siché non è miraculo che uno peregrino ingenio inpari uno di questi giochi facilmente. Con ciò sia che molte volte a me sia intervenuto, ma tu, candidissimo lectore, potrai, per varii et diversi modi coprire loro facilità. Per tanto non mi distenderò più che a volerti dare tucti i modi, saria troppo lungho et fastidioso el mio parlare. Basta che io ti ho decto el mio parere intorno a ciò. Hora tu seguirai quello che più ti comoda et più ti verrà per fantasia.

53v **Demostratio quarta a investigare tre parte de uno numero diviso**

Nota una altra bella et legiadra via da potere venire alla notitia di uno numero diviso in 3 parte, comme se dodici pure fussi diviso fra 3 persone, che uno ne havessi 3, l'altro 5, l'altro 4. Questo è uno modo dalli sopradecti differente, et fia questo dirai al primo ch'el suo numero lo multiplichi per 2, et all'altro dirai che multiplichi el suo numero per uno numero magiore ch'el numero diviso, cioè magiore di 12, non porta quanto sia, dalli la libertà, et al terzo, sempre per regola generale, dirai ch'el suo numero multiplichi per uno più ch'el secondo comme se el secondo multiplicha per 15, et lui multiplichi per 16. Et tucta questa somma, dirai che acozino poi insiemi et serbino. Et tu tacitamente da

2 ch'el numero] K ch'el il numero 8 quel] F questo 14 solo el modo] K el modo 28 da potere venire] F da venire 36–115.1 tu tacitamente da te] F et tu poi tacitamente

te multiplicha el numero diviso, cioè 12, per quanto facesti multiplicare al tertio. Et di questo dirai che cavino tucta la lor somma, et quello resta tucto da te partirai per 2 mancho che il numero per cui multiplichasti el numero diviso.
Hora ecchoti lo exemplo. Et sia el numero diviso 12 partito | in 3 persone, l'uno 54r
3, l'altro 5, l'altro 4. Dirai a quello che ha 3 che lo doppi et ha 6, et al'altro, cioè al secondo, che ha 5 dirai che multiplichi per uno numero magiore ch'el numero diviso, ché se lui multiplichassi per 1000 non porta. Hora diciamo multiplichi per 15, che è magiore ch'el numero diviso, farà 5 via 15, farà 75. Et al tertio dirai che multiplichi el suo numero per uno più che il secondo. Dunque dirai lo multiplichi per 16 farà 4 via 16, fa 64. Hora acoza tucto insieme, cioè 6 et 75 et 64, farà 145, et questo serbino. Hora tu da te tacite multiplicha el numero diviso, cioè 12, per quanto facesti multiplicare la terza parte, che la facesti multiplicare per 16. Dunque multiplicha 12 per 16, farà 192. Hora dirai che cavino la lor somma, cioè 145 di 192, resterà 47, quale partito per 2 mancho che il numero che fu multiplicato el numero diviso, cioè 2 mancho di 16, cioè 14, che partito 47 per 14 ne viene 3 per el numero che fu adoppiato, et avanza 5 per il numero secondo che fu multiplicato per uno numero magiore ch'el numero diviso per lo tertio numero. Acozerai 3 et 5, fa 8 che per insino in 12 vi manca
4 che | hai tucti 3 e' numeri. Et se tu havessi facto doppiare 5 fa 10, et l'altro 54v
multiplichato per 20, cioè 4, fa 80. Et l'altro multiplicato per uno più, cioè 3 per uno più di 20, cioè per 21, fa 63. Acozato tucto insiemi 10 et 80 et 63, fa 153, multiplicato di poi 12 numero, diviso per 21, cioè per il numero si multiplichò el tertio numero, farà 252, del quale tractone 153, cioè la somma loro, resta 99, quale partito per 2 mancho che il magiore multiplicatore, cioè per 19, verranne 5 per il numero che fu doppiato. Et avanzerà 4 per il numero che fu multiplichato per lo numero magiore del numero, diviso l'altra parte nel modo sopradecto, ti sarà noto la tertia parte. Et a ciò megliore et più mirabile possa fare tal giocho, ti voglio dare una regola generale che sempre farai questo giocho più bello et più stupendo nel modo sopradecto. Nota che per quello numero che tu fai multiplicare al primo, per tanto mancho che l'ultimo hai a partire. Et, perhò, nel sopradecto caso ti ho decto che tu parta el restante, cioè la differentia che è della tua somma alla sua per 2 manco | che la magior somma 55r
perché el primo multiplichò el suo numero per 2. Ma se el primo multiplichassi el suo numero per 3, per mancho, 3 havesti a partire. Et se multiplichassi per 4, per 4 mancho haresti a partire et se per 5, per 5 mancho partiresti. Hora nota, per più intelligentia, lo exemplo evidente. Sia partito 12 comme di sopra in 3 et in 5 et in 4. Dirai a quello che ha 4, multiplichi el suo numero per quanto vole, dandoli la libertà che parrà più bello. Hora diciamo multiplichi per 5, farà 4 via 5, fa 20. Hora al'altro che ha 5 dirai che multiplichi el suo numero per quanto vole, pure che sia magiore ch'el numero diviso. Hora sia per 24 multi-

plicato 5 24, farà 120. Et al'altro che ha 3 dirai che multiplichi per uno più ch'el
secondo, cioè per 25, farà 75. Acozino di poi tucti insiemi, havranno somma
che sarà 215. Hora tu da te multiplicha 12 via 25, farà 300. Et dirai ti dichino
la loro somma, overo che loro cavino la loro somma di 300 et dichino a te il
55v restante che te lo diranno perché omni uno saprà che questi giochi non se |
indovinano che tucto si fa per forza et via di multiplicationi et partimenti di
numero che niuno hoggi dì si trova indovino. Et perhò bisogna per qualche via
occulta venire alla notitia prefata. Dunque, tornando al proposito, cavando 215
di 300, resterà 85, la qual somma ti sarà nota. Hora dico che parta questo 85
per tanto mancho della somma multiplichasti el numero diviso per quanto tu
facesti multiplicare el primo numero che, se bene ti ricorda, lo facesti multi-
plicare per 5. Dunque cava 5 di 25 tuo multiplicatione, resterà 20. Et questo è
tuo partitore. Dunque partito 85 per 20 ne viene 4 per il numero del primo et
avanza 5 per el numero del secondo. El terzo sirà el restante per insino in 12
et se bene multiplichassi el primo numero per 20, diciamo multiplichi 3 per
20, farà 60 et poi multiplichi el secondo numero, cioè 5 per 25, farà 125 et il
terzo per uno più ch'el secondo per 26, farà 4 via 26 fa 104. Et acozato tucto
56r insiemi farà 289. Et tu, multiplicato 12 per 26, farà 312, et di | questo 312 cava
289, resta 23, quale partito per 20, mancho del tuo ultimo multiplicatore, cioè
di 26, hai a partire 23 per 6 viene 3 per il numero primo che fu multiplicato per
20 et avanzerà 5 per il secondo numero. Et serviratti sempre, ma nota che la
multiplicatione del primo numero debba essere minore che la multiplicatione
della seconda perché, altrimenti, non servirebbe la regola. Et perhò noterai et
terrai cura, dicendo quando lui ha multiplicato la prima volta, che multiplichi
per uno numero magiore che lla prima a suo modo, sebbene multiplichassi per
1000 riuscirà sempre. Et così potrai ritrovare el punto di 3 dadi o di dua tracti
pure che a te sia noto tucte le somme di tucti 3, ma non la loro distinctione.
Comme a dire: uno ha tracto 12 in 3 dadi sapere quanto ciaschuno habbi tracto
di per sé overo tre cose diverse le quali habbi tu batezate a numero, comme tre
grossi di 3 conii, cioè uno fiorentino, uno sanese et uno pesarino, ponendo a
ciaschuno el suo nome: uno dua, l'altro 3 l'altro 4, et in tucto hai 9, et poi divisi
56v fra | 3 persone, sapere chi ha l'uno et chi ha l'altro. Et molti vari et mirabili giochi
potrai fondare in su questa mia regola che a me saria fastidioso a volerla tucta
explicare ché empirei mille quaderni se tucti li volessi recitare. Ma allo spirito
gentile basta solo uno cenno et vienne mecho. Hora nota la seguente dimostra-
tione bella et gentile.

---

14 restante per insino] F resto insino 16 farà] F fa | numero, cioè] K numero, per cioè
19 resta] F resterà 28 12 in 3 dadi] F 12, 13 dadi 32 numerabili] F mirabili

Demostratio quinta a trovare 4 parte di uno numero diviso

Voglio darti modo a trovare 4 parte di uno numero diviso comme se tu havessi posto in tavola una quantità di grossi o quatrini et dicessi che quattro li dividessimo fra loro, cioè havessi posto 20 o 25 grossi in tavola, non inporta, et dicessi: "4 di voi li piglino et io mi vanto di trovare omni uno quanto ne habbi preso" che in verità sarà tenuta bella forza di numero. Et potrai fare in questo modo: cioè quando tu cavi fora ei grossi o quattrini, fa prima tu di haverne messo da canto una quantità a te nota, et poi quando tu | farai el giocho, non li havere a con- 57r
tare. Et fia tenuto dalla brighata più mirabile, ma questi giochi non si vogliono fare più che una volta perché sono tenuti più belli. Facendoli più che una volta sola perdono poi nel conspecto delle brighate. Sia dunque el numero diviso fra 4 persone 25. El primo habbi 4, el secondo 6, el tertio 7, et el quarto 8.

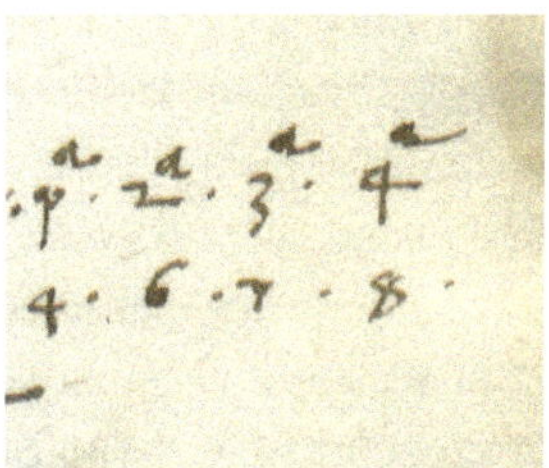

FIGURE 3
Marginal annotations on folio 57v

Et tu voglia investigare dicte parte a te incognite di che qualità sieno et chi lo tengha apresso di sé, terrà questa seguente regola generale che mai ti fallirà. Dirai che acozino la prima, la seconda et la tertia parte insiemi, et poi acozino la seconda, la tertia et la quarta. Et poi dirai acozino la tertia, la quarta et la prima insiemi et queste tre somme acozino insiemi tucte che, se bene notasti, sono 3 somme. Et poi dirai che serbino et tu multiplicha el numero diviso, cioè 25 per 3, et dirai che la somma loro la cavino della tua, et quello resta ti dichino. Et tu piglia quello restante et cavalo del numero diviso, cioè di 25, et quello resterà sarà la penultima parte, cioè la tertia parte. Eccho lo exemplo: acoza la prima, la seconda | et la tertia che hai 17, poi acoza la seconda, la tertia et la quarta, 57v
fa 21. Et poi acoza tertia, quarta et prima, farà 19. Hora acoza tucte tre queste somme insiemi, cioè 17, 21, 19, farà 57. Et queste caverai dalla multiplichatione del numero diviso in 3, cioè di 3 via 25, fa 75, dove cava 57 di 75, resta 18. Hora a te fia noto questo 18 et tu caverai 18 del numero diviso, cioè di 25, resterà 7 per la penultima parte, cioè per la tertia. Hora, havendo una parte per dua modi, potrai trovare le altre tre, cioè per la regola che io ti darò, overo per el modo che

---

11 dunque] F adunque 12 el secondo 6] K el secondo 5 20 resterà] F resta 23 fa 21] F avrai 21

nelle antedicte due demostrationi che ho ditto a trovare uno diviso in 3 parte che tu sai che li altri 3 che restano hanno 18 diviso in 3 parte. Hora piglierai el modo a te più comodo. Quando per questa via vorrai ritrovare le altre 3 parte, farai così: per trovare la quarta parte dirai che acozino insiemi la seconda et la tertia et la quarta parte. Et poi dirai che acozino la tertia et la quarta et la
58r prima. | Et poi dirai che acozino la quarta, la prima et la seconda, et tucte queste 3 somme acoza insiemi che hai per la prima somma 21, per la seconda 19, per la tertia 18. Fa somma tucto di 58, quale cavato del numero diviso multiplichato per 3, cioè di 75, resta 17. Hora cava 17 del numero diviso, cioè di 25, resta 8 per la quarta parte. Et per trovare la prima parte farai acozare la tertia et la quarta et la prima. Insiemi hai 19 et poi la quarta et poi la prima et la seconda, hai 18. Et poi la prima, la seconda et la tertia, hai 17. Et acozate queste 3 somme insiemi farà 54, quale cavato della multiplicatione del numero diviso in 3, cioè di 25, resterà 21. Hora cava 21 di 25 numero diviso, resta 4 per la prima parte. Hora hai 3 parte, cioè la tertia che è 7, et la quarta che è 8, et la prima che è 4. In tucto fanno 19. Hora guarda quello ti mancha insino in 25 et quello sarà la seconda parte che vi mancha 6. Et 6 dirai che sia la seconda parte et hai tucte le parte, et anchora potresti trovarla per l'ordinario comme acozza la quarta, prima et
58v seconda, hai 18. Et acoza prima, se|conda, tertia, hai 17, et seconda et tertia et quarta havrai 21. Acoza tucto insiemi, fa 56, cavato di 75 resterà 19. Cava 19 di 25, resta 6. Per la tertia parte questo è il modo, ma havendo 3 parte non bisogna afatigarsi a trovare la quarta, perché sarà el restante del numero diviso. Hora hai el modo bello et gentile et da te potrai trovare molte vie et modi di fare giuochi mirabilissimi.

Demostratio 6a a trovare 5 parte d'uno numero diviso

Quando uno numero fussi diviso fra 5 persone, cioè che tu havessi posto in su una tavola 25 grossi, a tuo modo non porta, et che tu dicessi a 5 persone che fra loro lo dividessino, terrai questo seguente modo: cioè diciamo che tu habbi diviso 25 fra 5 persone, cioè al'una tre, al'altra 4, al'altra 6, al'altra 7 et l'altra 5. Hora tu vorrai ritrovare quanto habbi ciascuno di per sé. Punctalmente terrai questo seguente modo. Cioè dirai che acozi insiemi la prima et la seconda et la
59r tertia et la quarta parte, harai 20. Et poi dirai che acozino la seconda et la tertia, | la quarta et la quinta, harai 22. Et poi dirai acozino la tertia, la quarta, la quinta et la prima, harai 21. Et poi dirai che acozino la quarta, la quinta, la prima, la seconda et harai 19. Hora dico che tucti questi numeri acozino insiemi, cioè

7 acoza] F acozino 9 resta] F resterà 14 resta] F resterà 17 che ti mancha 6 et 6 dirai che sia la seconda parte] F *omissis* 20 cava] F cavato 24 mirabilissimi] F mirabili 34 et poi dirai che acozino] F et poi acozino

20, 22, 21 et 19, farà tucto 82. Hora, perché tu hai 5 numeri, multiplicharai el numero diviso per 4, farà 100, del quale cavato 82, cioè la lor somma, resterà 18. Hora cava 18 del numero diviso, riestati 7 per lo penultimo numero, cioè per quello che ha 7 che è el quarto numero. Hora per trovare la figura seguente, cioè la quinta figura, potrai tenere dua modi, cioè el modo di sopra dato nella sopradecta demostratione, a trovare uno numero diviso in 4 parte, et tu sai ch'el numero di questa demostratione è 25, cavatone 7 della parte che hai trovata, ti resterà 18, diviso fra 4 persone, siché potrai tenere el modo a trovare 4 parte di uno numero. Et anchora potrai trovare le altre parte per questa regola, facendo così: dirai che acozino la seconda, la tertia, la quarta et la quinta, harai 22. Et poi dirai che acozino la tertia, la quarta, la quinta et la prima, harai 21. Et poi
aco|zino la quarta et la quinta et la prima et la seconda, harai 19. Et poi dirai 59v
che acozino la quinta et la prima et la seconda et la tertia, harai 18. Hora dirai che acozino insiemi tucte queste somme, farà 80. Et cavato 80 della multiplicatione del numero diviso in 4, cioè di 4 via 25, fa 100, resterà 20. Hora cava 20 del numero diviso, cioè di 25, resta 5 per la quinta parte. Hora hai 2 parte note, cioè 7 et 5, fa 12. Hora ti resta 3 parte che le poi ritrovare per la regola di uno numero partito in 3 parte. Ma volendola ritrovare per la regola presente, terrai questo modo: prima acoza la tertia, la quarta, la quinta et la prima, et hai 21. Poi acoza la quarta, la quinta, la prima et la seconda, hai 19. Et poi acoza la quinta, la prima, la seconda, la tertia, hai 18. Et poi acoza la prima, la seconda, la tertia, la quarta, farà 20. Et tucto acoza insiemi, harai 78 che cavato 78 di 100 resta 22, che cavato di 25 22, resta 3 per il numero primo, cioè per la prima parte. Et hai notitia di 3 parte. Resteratti 2, le quali potrai per più modi investigare, ma volendo investigare per questa regola, farai così: acozerai insiemi la quarta,
la quinta, la prima et la seconda, harai | 19. Et poi acoza la quinta, la prima, la 60r
seconda, la tertia, fa 18. Et poi acoza la prima, la seconda, la tertia, la quarta, harai 20. Et poi acoza la seconda, la tertia, la quarta, la quinta, harai 22. Et poi acoza tucto insiemi, harai 79, qual tracto di 100 resteratti 21. Et cavato 21 di 25 numero diviso, resterà 4 per la quarta parte che fia la seconda parte. Resteratti a ritrovare solamente la seconda, la quale ritrovarai per il modo sopradicto, ma per più brevità dirai sia el restante sino in 25, dove acoza 7 che tu trovasti, è 5 per la seconda, et 3 per la tertia, et 4 per la quarta, farà in tucto 19, insino in 25, resta 6. Et tanto dirai che sia la parte ti mancha a ritrovare. Et così potrai fare questo mirabile effecto et potrai dire alla brighata: "omni uno pigli una carta et io voglio sapere che carta lui ha presa". Potrai tu specificare overo porre in tavola 3 o 4 carte a te note comme 4 coppe, 6 spade, 7 bastoni et 3 denari, et

---

3 riestati] F riesteracti 14 80] F 8 16 resta] F resterà 17 resta] F resterà 23 resta] K resta resterà 30 resterà] F resteracti 31 il modo] F il numero

dire: "omni uno pigli di queste 4 carte quale vole", overo più bello potrai farlo a
torre i germini et porne 4 o 5, o 6. Et sebene volessi, tucti li potresti ritrovare con
60v questa regola, meglio e trionfi | che le carte, perché ei trionfi sono signati pun-
talmente puncto per puncto. Et potrai dire che non ti curi di vedere quali trionfi
loro piglino, ma che ti dichino tucta la somma quanto fa fra tucti. O se non te la
vorranno dire per li modi nella prima parte dati, potrai trovare la prima somma
che sarà più mirabile, dicendo: "io voglio indovinare quanto numero tu hai tra
tucti". Il che saputo et investigato, potrai trovare le sue parte divise nel modo
dato, ma a ttorre le carte non sapresti dire di che giuocho si fussi, potrai dire:
"tu hai 7", ma non le sapresti dire se è coppe, bastoni, spade o denari. Perhò è
meglio germini o trionfi che le carte. Potrai trovare puncti di 4 o 6 dadi et mille
altre cose belle. Et nota che se uno numero fussi diviso in 6 parte, ritroverale
in questo modo: sempre sia tua guida uno mancho che la parte divisa comme
nella superiore. Hai 5 parte et tu pigli 4 per guida, cioè multiplichi el numero
diviso per 4 et poi acozi insieme 4 parte cominciandoti dalla prima et seguendo
61r ordinariamente | comme nella sopradicta. Et s'el numero sarà dunque diviso in
6 et tu pigli 5, per guida, acozandovi insiemi la prima, la seconda, la tertia, la
quarta, la quinta et poi la seconda et la tertia, la quarta, la quinta, la sexta et
poi la tertia, la quarta, la quinta et sexta et prima et poi quarta, quinta, sexta
prima et seconda, et poi quinta, sexta prima, seconda, tertia, et acozando tucto
insieme et levando dalla multiplicatione del numero diviso in 5 et poi el resto
cava del numero diviso et harai la penultima parte, cioè la quinta parte. Et s'el
numero fussi diviso in 7 parte et tu piglierai per tua guida 6, et se sarà diviso in 8
piglierai per tua guida 7, et in 9 tu piglia 8, et così se bene fussi diviso in 100 parte
et tu pigla per tua guida 99. Et farai giuochi belli et amirandi nel conspecto di
molti, et operarai queste mie regole che ti serviranno benissimo a molte varie
cose che in questo libro non si recitano perché troppo lungo saria el mio trac-
tato. Et so che el vostro acutissimo ingenio, Magnifico Juliano, si extende a più
subtilità che a queste cose minime, ma io parlo sì aperto, perché qualche altro
61v so che harà lume di questo mio | libretto et perché el lectore gentile et da bene
possa senza troppa fatigha intendere questi secreti della scientia mathematica.
Hora andiamo innanzi.

Demostratio 7a a trovare uno numero pensato et non passi 105

Havendoti dato la superiore general regola a trovare quante parti tu volessi di uno numero, se bene fussi 1000 parte, non mi pare oportuno sopra simili casi stendermi più, ma voglio seguire qualche caso di numero terminato ad instru-

29 subtilità] F soctilità 34 havendoti dato la superiore general] F avendoti la superiora generale

irti. Comme dicessi uno tuo amico: “io ho uno numero che non passa 105”. Brevemente si potrà ritrovare et con una legiadra regola, overo che tu dirai a uno amico: “pigla uno numero che non passi 105 et io presto presto te lo voglio ritrovare”. Terrai questo modo: dirai che da 105 in giù pigli che numero si vole, che sia intero, perché essendo rotto la regola non servirebbe. Et questo numero, dirai lo parta per 3, et dicati solamente lo avanzo, non ti dica quello ne viene. Et poi lo parta per 5 et simili ti dica lo avanzo. Et poi lo parta per 7 et così ti dica lo avanzo. Et tu per generale regola sempre quando lui lo parte per 3, per omni | 62r unità che li avanza et tu tieni alle mani 70. Et quando lui lo parte per 5 per omni unità che li avanza, et tu tieni alle mani 21. Et quando lui lo parte per 7 per omni unità che li avanza, tu tieni alle mani 15. Et poi tucto quello che tu ti trovi alle mani et tu somma insiemi et questa somma poi la parti per 105, et quello ne viene non si cura, ma quello ch’avanza sarà el numero che l’amico tuo prese. Potrai dunque dire allo amico che partendo la tua somma per 105 li avanzerà apuncto el numero che lui prese. Comme se lo amico tuo havessi preso 31, hora dirai lo parta per 3, che li verrà 10 et avanzerà uno. Hora dirai non vuoi sapere lo advenimento et lo avanzo che non bisognerebbe investigare, altrimenti, ma solum ti dica lo avanzo che li resta, partendolo per 3 che, comme ho dicto, li avanzerà uno. Dunque per questo uno terrai alle mani 70 et poi dirai lo parta per 5, che partito 31 per 5 ne virrà 6 che non te l’ha a dire et avanzerà uno. Questo vorrai ti dica. Dunque, per questo uno tu hai alle mani 21 che posto sopra 70 fa 91. Et poi dirai lo parta per 7 | che partito 31 per 7 ne vieni 4 et avanza 3. Dunque 62v per 3 che ti avanza, hai a tenere alle mani 3 via 15, fa 45 che per omni unità hai a tenere alle mani 15. Dunque hai 3 unità, hai a tenere 3 volte 15, cioè 45, dunque posto 45 sopra 91, fa 136. Hora tu hai per tua somma magiore 136. Questa somma si debba partire per 105, verranne uno, che non si cura et avanza 31 che sia el numero pensato per lo amico. Et se lo amico havessi preso 44, che partito per 3 avanza 2, hai a tenere alle mani 2 via 70 fa 140, et partito per 5 avanza 4 che hai a tenere 4 via 21, fa 84. Et partito per 7 avanza 2 che hai a tenere 30. Acozato dunque 30, 84 et 140 farà 254 che partito per 105 ne viene 2 et avanza 44 per el numero che prese lo amico. Et nota che se lui ti dicessi: “io ho preso uno numero che partito per 3 et per 5 et per 7 non avanza nulla”, allora dirai essere el numero terminato, cioè 105, che sino a quello non si trova numero che partito per questi 3 numeri non avanzi almeno per uno, siché quando lui dirà non mi avanza nulla, dirai: “sia 105”. Et se lui | pigliassi 104, anchora nel modo pre- 63r decto, lo troverrai che 104 partito per tre avanza 2, et hai 140. Et partito per 5 avanza 4 et hai 84. Et partito per 7 avanza 6 che hai 90. Acoza tucto insiemi, fa

17 advenimento] F avenimento 30 ne viene] F ne verrà

314, partito 314 per 105 ne viene 2 et avanza 104 per il numero preso. Et sempre ti riuscirà questa regola. Et nota che quando lui havessi uno numero che partito per 3 et per 5 non avanzassi nulla et per 7 avanzassi qualche cosa, allora tu ti hai a tenere al numero che avanza. Comme se l'amico pigliassi 15 che partito per 3 avanza nulla et partito per 5 avanza nulla et partito per 7 avanza uno che per questo hai solamente alle mani 15, allora dirai quel tal numero essere 15. Et se l'amico pigliassi 21 che partito per 3 avanza nulla et partito per 5 avanza uno. Et hai alle mani 21 et partito per 7 avanza nulla. In questo caso hai a dire che el numero qual prese fu 21, perché questi dua sopradecti numeri non ascendano alla somma di 105. Et perhò non si partono per decto numero, perché ne avanzarebbe rotti et la regola nostra non presupone rotti. Comme se lui piglassi 70 63v che partito per 5 et | per 7 non avanza nulla et partito per 3 avanza uno solamente, che per questo uno hai a tenere alle mani 70. Et tanto dirai si ponessi l'amico. Et se lui si ponessi 35 che partito per 3 avanza 2, et hai alle mani 140, et poi partito per 5 et per 7 non avanza nulla, in questa hai a partire 140 per 105, ne viene uno et avanza 35. Et così ti riuscirà in tucti e' modi et serviratti la regola benissimo. Et potrai fare uno giocho bello et legiadro che senza domandare quasi niente ritroverrai omni numero che per lo amico fussi preso. Potrai ritrovare una carta, uno triumpho et simile o una quantità di danari posto in su una tavola.

Demostratio 8a a trovare uno numero che non passi 315

Se tu vorrai passare 105, dirai: "horsù io sono contento che tu passi 105, ma non passare 315". Et ritroveralo per una altra breve regola, comme potrai vedere, ma nota che quando tu entri nel giardino, cioè che tu fai questo giocho, bisogna che tu sia buono abachista, altrimente non vi ti mettere, perché questo giocho 64r vole bono cervello di sapere bene racorre et multiplichare senza | penna, cioè a mente. Et perhò, nota bene: quando l'amico harà preso uno numero et tu se' certificato che non passa 315, ritroveralo con questa brieve regola. Dirali che lo parta per 5, per 7 et per 9, et che di tucti questi numeri ti dica l'avanzo. Et tu noterai che per omni unità che li avanza quando lo parte per 5 et tu tieni alle mani 126. Et partendolo per 7 per omni unità che li avanza, tieni alle mani 225. Et quando lo parte per 9, per omni unità che li avanza, tieni alle mani 280. Et poi acozerai queste somme tucte insiemi et questo partirai per 315. Et quello ne viene, comme nella dimostratione superiore, non si cura, ma si nota quello

1 ne viene] F ne verrà 2 riuscirà questa] K riuscirà per questa 3–4 tu ti hai a tenere] F tu hai a tenere 18 ritroverrai] K li troverà 26 penna] F penare 29 per 5, per 7 et per 9] F per 5 et per 9

avanza. Et perhò parrà bello, dicendoli: “se tu parti la mia somma”, explicandoli quello che tu hai alle mani, “l’avanzo che resterà sarà el numero che tu piglasti”. O veramente tu li potrai dire explicatamente “tu piglasti tanto”, nominando el numero quale, operando comme di sopra, ti sarà noto. Hora sia che l’amico pigli 37 quale partito per 5 avanza 2, che hai a tenere alle mani. 2 via 126 fa 252. Et poi la parti per 7, | avanzerà 2 similiter et questi 2 terrai alle mani. 2 via 64v 225 fa 450, et poi lo parti per 9, avanzerà uno che per questo uno hai a tenere alle mani 280 apuncto. Hora dico che acozi tucti questi numeri insiemi, cioè 450, 252 et 280, faranno tucti 982. Et questo parti per 315, verranne 3 et non si cura, et avanzerà 37 che fia per il numero prese l’amico. Et comme nella superiore, se l’amico piglassi numero che partito per 5, per 7, et per 9 non avanzassi nulla, allora in quello caso dirai quel tal numero essere 315, perché niuno altro numero che insino a 315 si troverrà, che partito per li prenominati numeri non avanzi qualche cosa al mancho per uno di decti numeri. Et comme nella superiore, se lui pigliassi 35 che partito per 5, per 7, non avanza nulla et partito per 9 avanza 8, questo avanzo ti servirà solo perché piglia 8 volte 280, sarà 2240. Et questo parti per 315, ne verrà 7 che non si cura, et avanzerà 35 per il numero che prese l’amico. Et se lui piglassi 45 che partito per 5 non avanza nulla, et partito per 7 avanza 3, che per questo 3 hai a tenere alle mani 3 | via 225 fa 675. Et 65r questo parti per 315, ne verrà 2 et avanza 45 per il numero prese lo amico. Et se lui pigliassi 63, che partito per 5 avanza 3, et hai alle mani 3 via 126 fa 378. Et partito per 7 avanza nulla, et partito per 9 avanza nulla. Et tu partirai 378 per 315, verranne uno et avanzerà 63 per lo numero prese. Et se lui pensassi 4 che non si pò partire per niuno delli prenominati numeri, in questo caso sempre dirai che avanza 4 et a te sarà noto che lui prese 4, perché non è possibile che uno numero da 315 in giù, partito per li prefacti numeri, sempre avanzi 4. Et, perhò, senza afatigare altrimente el tuo cervello, dirai che piglassi 4. Et questa regola è infallibile. Et così farrai nella superiore di 105 che quando pigliassi 2 che non si può partire per niuno delli sua partitori, allora in quel caso dirai sia el suo numero preso: 2, perché, comme è decto per insino, in 105 non si trova numero che partito per li sopradecti partitori, cioè per 5, 7, et 9, sempre avanzi uno medesimo numero, comme tu da te potrai provando vedere, ché così | potrai ritrovare omni quantità che fussi posta in tavola, pure che tu sia 65v certificato che non passi dicto numero 315. Et quando l’amico piglassi numero in quore, specificale che non pigli rotti, perché la regola non ti servirà, sendovi rotti, né in questa né in la superiore.

16 ti servirà] F ti seservirà 17 ne verrà] F verranne 30 sia el suo numero preso 2] F sia el suo numero 2, cioè el numero preso

Demostratio 9a a trovare uno numero partito per 2, 3, 4, 5, 6, avanza uno et partito per 7 avanza nulla

Hora voglio porre alcuni casi quali, benché sieno noti quasi al vulgo, pure occorre alle volte che una persona, quale ha moltissime occupationi comme ha Vostra Illustrissima Magnificentia Magnifico Juliano, non può simile cose minime sempre havere a memoria. Et a ciò quella in questo mio volume habbi tucti questi simili casi, perhò mi sono disposto recitarli in questo mio libretto. Et per essere molti noti, mi sforzerò recitarli sotto brevità che non vi fia mirabile se fra qualche rose che sarà qui inclusa ancora qualche spino, bisogna recitare anche questi. Et chi ne ha notitia li preterirà, saranno scripti per quelli non li
66r sanno | et non hanno lume. Dunque diremo così: se'l fussi uno che ti domandassi: "trovami uno numero che partito per 2 et per 3, per 4, 5 et 6, avanzi sempre uno, et partito per 7 avanzi nulla", a questi poi rispondere che se ne trovano infiniti che non si può dire: è il tale. Ma se tu dicessi quale è il primo, quale è secondo, è tertio che facci questo effecto, a questo dico che si potrà rispondere. Et dirai quale è il primo, quale è il 2, etc., che il primo che farà tali effecti è 301, el secondo è 421, etc. Ma voglio dare regola generale in questo. Tu hai a trovare numero che habbi 1/2, 1/3, 1/4, 1/5, 1/6, che el primo che habbi tucte queste parti è 60, dove, partito 60 per tucti li prenominati numeri, non avanzerà nulla. Posto, dunque, uno in su 60, farà 61 che partito per tucti e' decti numeri sempre avanzerà uno. Hora partito 61 per 7, avanzerà 5. Et tu vorresti avanzassi nulla. Dunque dirai: "io ho 60 che partito per tucti lo numeri mi serve, ma io vorrei che partito per 7 avanzassi 6, et 61 partito per 7 avanza 5". Dunque multiplicharai sempre quello tale per quello che avanza, cioè per 5, 5 via 60 fa 300.
66v Et questo | partito per 2 et per 3, 4, 5, 6, sempre avanza nulla. Et partito per 7 avanza 6. Dunque porrai uno in su 300, fa 301, et questo farà el bisogno di tucto. Et se più ne volessi, potrai trovare infiniti, ma nota che la regola di sopra data non è generale, per non parlare fora di ragione, perché 120 partito 2, 3, 4, 5, 6, avanza nulla che se partito per 7 avanzassi 6 faria el bisogno, ma 120 partito per 7 avanza uno. Ecco che non fa el bisogno. Ma regola generale sarà a dire "trovami uno numero che habbi 1/2, 1/3, 1/4, 1/5, 1/6", che lo troverai multiplicando le figure di sotto insiemi, cioè 2 via 3 fa 6, et 4 via 6 fa 24 et 5 via 24 fa 120, et 6 via 120 fa 720. A questo giugni uno, fa 721. Et questo farà el bisogno. Et da te col tuo ingegnio ne potrai trovare infiniti. Et se tu porrai sopra 301, 300 farà 601. Et questo anchora farà el bisogno. Et poi se tu porrai sopra 601 300, fa 901, et questo sarà 901, farà el medesimo. Et se più ne volessi, potresti in infinito pro-

3 voglio porre alcuni] F voglio alcuni 8 vi fia] F fia 13 trovano] F truova 14 può] F po' 22 tucti] F tucto 23 avanza 5] F avanza 6

cedere, ponendo sempre 300 sopra el numero et sempre ti riuscirà. Et questa sia regola generale. | 67r Et questo basti quanto a questa dimostratione. Hora nota un'altra quasi simile alla precedente.

Demostratio xa a trovare uno numero che partito per 2 avanza 3 et partito per 3 avanza 2 et partito per 4 avanza 3 et partito per 5 avanza 4 et per 6 avanza 5 et per 7 avanza nulla

Una altra forza si dimostra nel numero. Comme a dire: s'el tuo amico dicessi o pensassi uno numero diviso in 2 avanza uno, et diviso in 3 avanza 2, et diviso in 4 avanza 3, et diviso in 5 avanza 4, et diviso in 6 avanza 5, et diviso in 7 avanza nulla, vorrei lo indovinassi che numero è questo. A questo respondemi che molti numeri et quasi infiniti sono quelli li quali fanno tali effecto, ma volendo sapere quale è el primo, troveralo in questa forma. Trova uno numero che habbi 1/2, 1/3, 1/4, 1/5, 1/6 che lo troverai dicendo: 2 via 3 fa 6 et 4 via 6 fa 24 et 5 via 24 fa 120 et 6 via 120 fa 720, ma troverrai anchora 60 averà tucte decte parti. Hora dirai 60, cavane uno, resterà 59, et questo 59 partito per tucti li prenominati numeri. | 67v Sempre avanzerà uno mancho che il partitore dove se partito per 7 non avanzassi nulla ti servirebbe, ma perché partito per 7 avanza 3 non fa el bisogno. Hora, a volerlo ritrovare poni 60 in su 59, farà 119. Et questo farà el bisogno per tucti li prenominati numeri. Dunque dirai che 119 sia el primo numero che partito per il modo sopradecto avanzerà sempre uno mancho del partitore, et per 7 avanzerà nulla. Potrane trovare più, ponendo sopra 119 questo numero, cioè 420, farà 539. Questo farà el medesimo effecto et se più ne vorrai, porrai sopra 539 una altra volta 420, farà 959. Et questo farà anchora el bisogno. Et così potrai prendere in infinto, ponendo sempre in sul numero che hai 420 et haranne infiniti comme per te potrai vedere. Et se tu vorrai uno numero che partito per 2 avanzi uno, et partito per 3 avanzi 2, et partito per 4 avanzi 3, et partito per 5 avanzi 4, et partito per 6 avanzi 5, et partito per 7 avanzi 6, et partito per 8 avanzi 7, et partito per 9 avanzi 8, et partito per 10 avanzi 9, et per 11 venga apunto, terrai questo modo comme nella superiore. | 68r Trova uno numero che habbi 1/2, 1/3, 1/4, 1/5, 1/6, 1/7, 1/8, 1/9 che tu sai che 60 ha 1/2, 1/3,1/4, 1/5, 1/6. Hora, perché habbi 1/7, multiplicherai per 7, farà 420. Et perché habbi 1/8 lo multiplicharesti per 8, ma perché ha 1/4 perhò basta solo che lo multiplichi per 2, cioè nel ripiego di 2 via 4 che fa 8. Et perché lo multiplichasti per 4, basterà che hora lo multiplichi per 2, farà 2 via 420, farà 840. Et perché habbi uno 1/9, bisognerebbe lo multiplichassi per 9, ma perché già l'hai multiplicato per 3, lo multiplicharai una altra volta per 3 et sarà multiplichato per 9, onde multiplichato 840 per 3, farà 2520, dal quale caverai uno, resterà 2519. Et questo farà li

---

13 due via tre fa] F *omissis*

prenominati effecti, ché partito questo 2519 per tucti li numeri fino a 11 sempre avanzerà uno, mancho che il partitore ripartito per 11 avanzerà niente. Hora, volendone più di decti numeri, porrai 2520 tante volte sopra questo numero che facci el bisogno. Et se tu vorrai più numeri, che faccino questi effecti, col tuo mirabile ingenio potrai trovare infiniti et per tanto in queste cose note al vulgo non mi distenderò più.

68v **Demostratio XIa a indovinare una conpera che uno facessi senza alcuna multiplichatione**

Sarà caso bello, benché sia in sé facile; pure chi non intende sua facilità, li parrà admirando. Ma nota che questi giochi facili a comprenderli non si vogliono fare per più che una volta, perché facendoli, poi più facilmente si possono intendere. Hora nota: dirai a uno che pigli di dua sorte monete a te note et non ti dica la quantità comme "io ne piglio tante", ma te dica la qualità delle monete: "io piglio grossi, quatrini et piccioli etc." Et che di una di decte monete comperi pere o pesche a suo modo et dicati solamente d'una di decte monete quante pere comperò. Et tu ti oblígha comparando poi, lui delle altre monete, alla medesima ragione, sapere dire quante pere compera la seconda volta, ché a chi non harà così vera notitia de sua facilità, sarà tenuto caso bello. Hora nota, farai così: dirai allo amicho tuo che pigli dua sorte | monete, cioè piccioli et quatrini che omni uno sa che di piccioli ne va 4 al quatrino nero, ma in qualche altro luogho 5. Hora diciamo che lui pigli quatrini neri et piccioli, che ne pigli tanto del'uno quanto del'altro, ché così per la regola nostra che altrimente non ti servirà la regola. Hora diciamo lui habbi preso 6 piccioli et 6 quatrini, la quale somma a te sia incognita, ma sai bene che una sorta è piccioli, l'altra è quatrini. Hora dirai che conperi di piccioli tante pere o pesche a suo modo batezerala tu et dicati de' piccioli che lui ha, quante pere che lui ha conperate et quante lui ti dice havere comperate la prima volta. Et tu multiplicha per 4 et tante ne conpera la seconda volta, verbi gratia, sia che lui dica havere conperato con e' denari che lui ha 3 pere. Hora tu dirai: "3 via 4 fa 12", et tante ne conperò tra tucti la seconda volta. Et acozate con la 3 prima dirai: "15 pere conperò fra tucte due le volte". Et per darti generale regola a questo, nota che diciamo lui de' piccioli havere conperato 3 pere, hai a conside|rare che omni quatrino vale 4 piccioli. Dunque, havendo lui tanti quatrini quanto picioli, verrà la seconda volta a conperare 4 volte tante pere che non conperò la prima. Et se lui pigliassi

(margin: 69r at "farai così"; 69v at "de' piccioli havere conperato")

9 Sarà] K V'era, F ara (con iniziale mancante) 10 non si vogliono fare] F non bisogna fare 13 Comme: io] F comme dire io | dica] K dico 15 pesche] K, F pesce 16 comparando] F comperdere 23 servirà la regola] F servirà 27 havere] K haverne 28 havere] K haverne 29–30 ne conperò tra tucti] K ne conperò 31 per darti] F et darocti

12 piccioli et 12 quatrini a te incogniti et la prima volta conperassi octo pere, la seconda volta ne conperassi 4 via 8, fa 32, et 8 prima fa 40. Dunque dirà che fra la prima et la seconda ne conperassi 40. Et meglio coprirai la facilità sua, dicendo quello conpera in tucto che dirli la seconda di per sé. Et se lui volessi conperare prima de' quatrini et poi la seconda di piccioli, dirai che a questo se' contento, dicendoti lui la prima volta havere conperato di quatrini, lui ha preso 12 pere. In questo caso hai a partire le pere per la valuta della minore moneta nella magiore, comme el quatrino vale 4 piccioli. Dunque parti 12 pere per 4, ne verrà 3. Et la seconda volta conpera 3 pere. Et se lui dicessi: "io voglio conperare la prima volta a mio modo et non lo voglio dire, ma ti dirò la seconda", et tu medesimamente potrai fare comme se lui havessi | preso 6 quatrini et tanti piccioli 70r
et conperassi la prima volta a suo modo et spendessi piccioli et la seconda conperassi con quatrini 12 pere et allora tu partirai 12 pere per 4, verranne 3. Et harai la notitia et lume della prima conpera. Et se lo amico pigliassi grossi et quatrini, cioè 6 grossi et 6 quatrini a te incogniti, et che prima spendessi e' quatrini et conperassi 4 pere, la prima volta dico che tu multiplichi 4 via 21, fa 84, perché el grosso vale 21 quatrino. Et perhò si multiplicha per 21 et dirai che la seconda volta conperò 84 pere. Et così quando piglassi ducati o carlini o iulii o bolognini, sempre la seconda volta, o tu multiplicha, o tu parti, secondo bisogna, per la valuta della minore moneta nella magiore et harai notitia vera, etc.

Demostratio XIIa a fare sempre uno numero innanzi al conpagno o qualunche altro numero. Regola generale, cioè 30

Una legiadra forza di numero si dimostra nel giuchare al 30 con uno dado comme molte volte ho giuchato io et vincto. Et dove si giocha a sorte voglio dia libertà | allo amico chiedere che puncto di dado lui vole. Et servando la regola 70v
qui seguente, non potrai mai preterire che sempre farai 30 o qualunché altro numero innanzi al conpagno, ponendo uno sopra l'altro sigillatim. Farai così: parti 30 o per 7 et quello avanza pigla tu, dicendo: "io piglio dua" et tu pigli lui quanto vole tu agiugnerai a 9 et lui pigli quello vole, tu agiugnerai a 16 et pigli quello vole et tu agiugnerai a 23. Et poi pigli lui quello vole, tu farai 30 avanti lui et questo nasce per la forza del numero ch'el dado sempre fa di sotto et di sopra. Se bene noti, sua proportione ha 7 puncti. Et perhò, sempre tu piglerai el numero che è sempre sotto al suo, di modo che fia sempre el puncto che pigla lui et il puncto che pigli tu farà 7 apuncto, onde ci è il vantaggio di chi pigla prima. Et nelli numeri che partiti per 7 avanza nulla, el pigliare prima è

---

23 si dimostra] F si mostra7 29 et lui pigli] F et pigli lui 34 che pigli] K che tu pigli
34–35 vantaggio di chi pigla] F vantaggio dico pigla

disavantaggio, ma nelli numeri che partiti per 7 avanza qualche unità comme 71r 30 partito per 7, avanzerà 2, dico che in questo, cioè a giuchare | al 30, è vantaggio grandissimo a uno pigliare prima. Et perhò, quando sono dua che omni uno di loro sa la malitia, sempre chi piglerà prima vincerà, perché pigli poi lui quanto vole che non può pigliare più che 6 et tu piglerai uno di modo che fra el puncto tuo et suo sarà sempre 7. Et per forza sempre farai 30 innanzi a llui. Ma quando el numero fussi 35 o 42 o 70 et simili, che partito per 7 avanza nulla, in questo caso dico che ha vantaggio colui che è il secondo a pigliare, perché havendo a fare per 7 sempre se tu piglassi uno lui pigliarebbe 6 et faria 7. Et così di mano in mano. Et perhò, nelli numeri partiti per 7 avanza niente, lasserai sempre pigliare al conpagno et nelli numeri che avanza qualche unità, comme a fare 50, tu partirai 50 per 7, avanzerà uno, dunque tu piglierai questo uno et poi pigli lui quello vole, sempre tu farai 50 et lui non, perché sempre piglerai tanto che fra el tuo puncto et 'l suo sia 7. Et così facto, farai el tuo intento, ma se lui dicessi: “io non voglio pigliare puncto di dado perché havessi cognosciu- 71v to che tu vi hai su malitia”, | in questo caso dirai essere contento che mecta insino a quanto si possa pigliare el magiore puncto, comme a dire facciamo a non passare a pigliare 8, ovvero a non passare 10, ovvero non passare 12, perché bisogna dare puncto terminato ché se tu dicessi a chi prima fa 50 et non dessi puncto terminato, el primo che piglia direbbe 50 et harebbelo facto innanzi a te. Et perhò bisogna dare uno termine di puncto, cioè dire a non potere pigliare più che 8 o 9 o 10 o 20, etc. Et tu per magiore suo scharico perché non dica che tu vi habbi malitia, dirai che termini a suo modo et lui, dandoli tu la libertà, non penserà che tu vi possa porre su malitia. Hora diciamo che lui dica: “io voglio a chi prima fa 105 et non passare 9 a pigliare”, cioè che niuno, dirai, possa pigliare più che 9 per volta, ma possa pigliare 9 o mancho a suo beneplacito. Hora noterai questa generale regola che servirà in tucti e' modi che tu potessi o volessi giuchare in questo modo: sempre piglia el numero terminato, cioè el magior numero che tu puoi pigliare, et in su questo poni uno et poi 72r parti | el numero che tu vuoi, farà innanzi al conpagno per il numero terminato, con quello uno tu ponesti su che in questo caso el numero terminato è 9 che non si pò pigliare più di 9. Hora dico che in su questo 9 ponga uno, fa 10. Et questo sarà tua guida. Prima parti el numero che voi havete a fare, cioè 105, per questo 10, che ne verrà 10 che non fa nulla, ma che avanza 5, el quale sarà tua guida a fare 105 sempre avanti al conpagno. Et perhò, havendo a pigliare, tu piglerai questo 5 et poi pigli el compagno che numero si vole. Et tu piglierai tanto che sia el numero che piglia lui et tu facti apuncto 10 comme tu hai preso

4 pigli poi lui] F pigli poi l'altro   30–31 terminato con quello uno tu ponesti] F terminato et con quello uno più che tu vi ponesti   32 più di 9] F più che 9

5, et lui piglerà 6, farà 11. Et tu piglerai 4 che farà 15. Et hai preso fra tu et lui 6, et 4, 10 et havete 15. Hora lui pigli 9 et tu uno et havvete 25. Et lui 7 et tu 3 fa 35. Et lui 5 et tu 5, farà 45. Et lui 3 et tu 7, fa 55. Et lui 4 et tu 6, fa 65. Et lui 6 et tu 4, fa 75. Et lui 8 et tu 2, fa 85. Et lui uno et tu 9, fa 95. Hora pigli lui quello vole et tu farai 105 innanzi a llui et harai vinto. Et per questa regola potrai omni | et qualunche numero fare innanzi al conpagno, ma quando lui 72v havessi a pigliare et epso sia experto, non potrai mai vincere, ma non essendo bene experto della regola, et benché lui di principio habbi a pigliare, lo potrai corre alla rete a uno passo comme aspectarlo a 25 o a 35, che spesso a me è intervenuto, ma potrai porre questo partito che non fia da recusare. Dirali che pigli el numero terminato et che se' contento che lui pigli prima, ma che tu vuoi porre el numero a che voi volete ascendere et per sua satisfactione porrali 2 o 3 numeri innanzi, acomodati acciò videlicet che questi numeri partiti per uno più ch'el numero terminato non avanzi nulla. Et a questo modo non potrai perdere et sempre vincerai comme in questo modo passare 9. Dirali che se' contento che lui pigli prima, ma che vuoi fare a chi prima giugne a 100 o 80 o 60, quale lui vole, che essendo contento a uno di questi, pigli lui che vole, tu la prima volta farai 10 et poi la seconda 20, et poi 30, et 40 et 50 et 60, etc. Ma nota che con quelli quali hanno lume di | questa regola non è vantaggio a giuchare, 73r ma io ho trovato di quelli che hanno lume della regola del dado, solamente poi cavandoli di quella li ho vinti, siché la experientia in sul facto t'insegnerà molti et vari colpi che saria prolisso el volerli tucti recitare, ma questi bastino in quanto a questo.

Demostratio XIIIa a giuchare al giuocho della morra et non perdere mai servando certe conditioni, etc.

Voglio dare regola generale a giuchare al giuocho delle corna per più legiadro nome decto "alla morra" et non perdere mai che fia bello caso. Et fia in questo modo: dirai a uno tuo amico o con qualunche tu volessi a questo giocho giocare che li voi mettere el seguente partito che tu vuoi giucare et in questo modo che tu vuoi tenere li occhi chiusi et non vedere niente, ma bene voli udire et che voli che lui levi innanzi a te et dica alta voce quello che lui chiede, ma solo vuoi che non possa chiedere el tuo puncto, cioè uno puncto che tu ti vuoi reservare che quando vorrà giucharli ne dirai. Et che tu | vuoi fare alla mana, cioè chi fa prima 73v 5 colpi et a llui di 5 ne vuoi dare vantaggio 4, cioè che vuoi havere tu a vincere 5 poste et lui habbi a vincere solamente una posta. Et per più largità porrai el partito in libertà dicendo: "o vuoi dare o vuoi torre a chi dico" ché uno che non sia di

2 havvete] F faze 9 che spesso] F come spesso 28 tu volessi a questo giocho giocare] F tu volessi giugare a questo giugo 33 chi fa] F a chi fa 34–35 5 poste] F 4 poste

questa forza experto piglerà sempre el vantaggio ché non noterà la malitia vi è sotto et piglerà stare a occhi aperti. Ma tu, stando a occhi chiusi, non potrai mai perdere tenendo la seguente regola. Ma comme nella superiore dimostratione ti dissi, con quelli che di simili cose sono experti non giuchare se tu non vedi el giuco vincto et netto perché è poi peggio la vergognia ch'el danno che io per me non vorrei essere vincto per niente ne' casi che io mi sento sperto ché, comme ho decto, non stimerei el perdere se non vi fussa dentro vergogna. Et perhò starrai in te et volendo vincere farai così: quando voi siate fermi, ovvero quando lui
74r volessi che tu li dicessi el numero che tu vuoi per te, dirali "io | voglio 6 et non voglio che tu possa chiamare 6, ma tucti li altri numeri ti do in tua libertà che tu possa chiamare quanto voli". Ma perché non pigli malitia dirai: "io non voglio che tu chiami el mio numero perché scontrandoci insiemi a chiamare 6 et e' fussi, harai perduto perché tu leveresti uno et havresti 5, et io harei uno solamente, dunque non farebbe per me. Et perhò, voglio che tu ti obblighi a non potere chiamare 6, ma tucti li altri metto in tua libertà". Et tu per più dimostratione non ti obligare a chiamare sempre 6, ma che vuoi potere chiamare che numero a te pare et 6 a tua posta. Et bisogna che tu habbi uno fidato che ti faccia el tuo giuoco. Et quando tu vieni alla conclusione che tu vorrai giuchare, tieni questa regola comme è dicto. Lui chiamerà alta voce ché tu senta quello che lui chiede et tu con la tua morra leverai numero che lui non possa fare el suo chiesto ché tu sai che lui non può levare più che 5. Et chiedendo 10 et tu leverai 4 o manco che non potrà fare 10. Et tu chiedi numero che tu credi che
74v sia per venire perché chiedendo sempre 6 | l'amico penserebbe malitia presto et perhò a te basta tenere lui ché mai non possa vincere tanto che tu faccia le tue 5 poste. Et perhò andrai variando le tue chieste, et se lui chiedessi 9 et tu alzerai 3 o mancho, che lui non potrà fare el suo 9. Et se lui chiedessi 8 et tu alzerai 2 o mancho, et non potrai lui fare 8. Et se epso chiedessi 7 et tu alzerai uno et non potrà fare 7. Et se lui chiederà 5 et tu alzerai 5 et non potrà fare 5. Et se lui chieda 4 et tu alza 4 o 5. Et se lui chiede 3 et tu alza 3 o più a tuo modo. Et se lui chiede 2 et tu alza 2 o più, ecco che lui chiegga quanto vuole non pò mai vincere tenendo tu el modo sopradicto. Et stando bene in cervello che se tu eri una volta sola, tu perderai, siché starai in te et non potrà fare che qualche volta tu non lo giunga. Et sempre vincerai servando questo ordine, ma vedi di fare con una certa maniera et gratia che il conpagno non si acorgha
75r della astutia che tu hai introclusa, la qual cosa non si pò insegnare. | Ma quando lo amico volessi mettere anchor lui conditione, ovvero niuno possa alzare più che 4, ovvero niuno possa alzare più che 3, che ponendo questa legge anchora

---

6 sperto] F experto 12 el mio numero] F el mio punto 21 tu sai] F tussai 22–23 che sia per venire] F che possa venire

potrai vincere faciendo così: senpre pigla el numero che tu vuoi che lo amico non possa chiedere, sia più uno che quello poni per legge che si possa alzare. Comme se ponessi per legge che non potrà alzare più che 4 per uno, in questo caso piglerai per tuo numero 5 et alzi lui 4 et chiega 8. Et tu, sapendo che non può alzare più che 4, alzerà 3. Et se lui chiede 7 et tu alza 2. Et se lui 6 et tu alza uno. Et così mai non perderai et quando facessi con tucta due le mani comme ho visto fare che è giuocho bello. Et tu nota che niuno può alzare più che 10, dunque piglerai 11 per tua guida che non potrà mai l'amico vincere che se lui potessi chiedere el tuo 6, ovvero comme nella seconda terminata, ovvero 11, a te saria inpossibile potere vincere mai, ma tenendo 11 per te che mai lo possa chiedere lui. Et tu starai vigilante sempre a quanto lui chiede alzando in modo che lui non | possa vincere che se lui chiede 20, 19, 18, etc. et tu alza 1,2,3, etc. Et 75v
se lui chiede 10 o 9 o 8 et tu alza 10 o 9 o 8, etc. Et questo basti.

Demostratio xiiiia a giuchare al tocho a non perdere mai servando certe conditioni

Anchora nel giuocho decto toccho in lingua fiorentina, ovvero in qualche luogho dicto conto, vi è dentro qualche astutia, le quali quando saranno servate da uno, mai potrà perdere. Ma bisogna trovare uno che giuochi con techo perché le brigate hoggi dì mi paiano tante astute et in omni malitia experte che poco si pò con epse guadagnare. Pure se tu harai notitia di tucti i giuochi et forze di questo mio libretto, non potrà fare che alle volte non colgha qualche una et perhò non ti spiaccia a llegere quando tu te trovassi in conpagnia con 3 o 4 o 6 conpagni comme si fa et che fussi a pagare qualche cosa comme fussi pippioni o uno par di polli et simili et vollessi pagare a toccho, cioè a chi toccha paghi pippioni o polli et tu non voglia perdere | dirai: "horsù facciamo questo 76v
pacto che omni punto che si alza vaglia tanto", explicando el numero comme di sotto adesso dico. Et prima se voi fusse dua, dirai che vuoi essere toccho tu, et che omni punto che si alza dica 2. Et mai perderai che se fra tucti a dua levassi 3. Et facendo ordinariamente perderesti, perché el numero caffo sempre toccha a te. Et facendo che omni uno dica 2, farà 6 et sarà numero pari. Et el numero pari toccherà sempre al tuo conpagno. Et potrai anchora dire omni uno vaglia 4 che fia tanto a dire omni vaglia 4 quanto omni uno vaglia 2. Et sia quello medesimo. Ma s'el compagno volessi essere toccho lui, allora tu dirai che alzando puncto

17 vi è dentro qualche astutia] F vi è drento qualche malitia et astutia 19 hoggi dì mi paiano tante astute] F mi paiono hogi sì tanto astute 23–24 comme fussi pippioni o uno par di polli et simili] F come uno paio di polli o pipioni o simili 25 et tu non voglia perdere] F et tu non voglia perdere, farai in questo modo 32 tanto a dire omni vaglia 4] F tanto che a dire omni uno vaglia 4 33 che alzando puncto] F che omni puncto che si alza

pari vuoi che li se giungha 5 o 7 o 9 o altro punto che sia caffo, purché el numero diventi caffo perché el numero caffo sempre termina in lui. Alzando caffo non bisogna porvi legge. Ma quando voi fusse 3 a fare al toccho, patteggierai che omni puncto che si alza vaglia 3. Et quando fussi 4, dirai vaglia 4. Et quando
76r fussi 5 vaglia 5. Et se | 6, vaglia 6. Et così in infinito, et mai toccherà a te pagare. Et sempre facendo così, toccherà a quello che è a llato a te. Hora tu comoda o a mano dextra o a mano sinistra, comme tu vuoi et secondo a chi vuoi che tocchi. Se vorrai che tocchi a quello che sta a mano ritta, va a mano mancha. Et se vorrai che tocchi a quello che sta a mano mancha, et tu va a mano ritta. Et sempre riuscirà. Ma comme è dicto, bisogna fare con certa maniera et gratia, ché le brigate che sono astute non se ne accorghino del facto, ché non ti riuscirebbe, siché con tuo mirabile ingenio potrai fare, perhò nota.

Demostratio XV a giuchare a pari o caffo et non perdere servando certa conditione

Possi anchora giuchare al giuocho dicto pari et caffo et fare in certo modo et partito che mai perderai mectendovi certe conditioni. Et fassi in questo modo. Dirai: "io voglio giuchare techo a non ingannare et, a causa che fra noi non
77r habbia essere differentia, voglio giuch|are col testimonio" perché molte volte io piglio et dico: "chiedi" et lui dice: "caffo". Et poi quando vede havere perduto, dice: "non dissi caffo, ma dissi pari". Per il che molte volte sono venuto in questione et spesso mi è incontrato havere poi havuto a fare alla pugna et per questo fugire, volendo giuchare a questo giocho, voglio fare col testimonio. Onde, sotto questa coverta di non volere havere a fare questione, forse indurrai el conpagno a giuchare comme più volte ho io inducto qualche uno. Et, volendo el conpagno giuchare, tu terrai questo ordine; dirai: "io voglio che comme io, ovvero tu hai preso et che io li dico verbi gratia caffo, che si tolgha una del numero delle prese et ponghasi da parte per testimonio che io ho decto caffo. Et se io dico pari, che si tolgha dua delle prese, mectasi pure da parte in testimonio che io ho dicto pari. Et questo testimonio non voglio poi si conti nel numero, cioè che stia là da parte solo per quello effecto et non si habbi a contare nel pari, nel caffo". Et in questo caso, omni volta che viene che tu habbi a pigliare
77v tu et lui si hab|bi a porre, non perderai mai più. Et perhò ingegnati di havere a pigliare la prima volta che mai li tornerà in mano. Et se pure lui li tocchassi a pigliare prima, non può fare che in 3 o in 4 poste tu non vincha una. Et metti pocho alla volta se lui già non havesse questa malitia che, havendola, non te ne fidare, ché bene presto te ne acorgerai se lui ha la malitia o sì o non. Et quando

20 per il che molte volte] F onde molte volte poi 24 et volendo] F onde volendo 26 ovvero] F ho 35 havendola] K havendolo

tu hai a pigliare, sempre per regola generale piglia in caffo et in modo che lo amico non si accorgha che tu anoveri omni volta. Et pigliando caffo mai potrai perdere, perché se tu hai caffo in mano et lui chieggha caffo, tu porrai caffo da canto per testimonio et in mano ti rimane pari. Et se lui chiede pari et tu hai caffo in mano, porrai pari da canto per testimonio et similmente ti rimane caffo in mano. Eccho che sempre vincerai, pigliando caffo. Hora farai tu con dextro modo.

Demostratio XVIa al giuocho di 15 giudei et 15 cristiani

Benché questa favola sia hormai molto | divulgata, pure non voglio preterirla 78r
che in questa mia opera non sia brevemente recitata che trovandosi in una nave XV giudei et 15 christiani et venendo fortuna grandissima, fu dal patrone della nave exposto che se la metà di loro non si gittavano in mare, tucti perirebbono. Dove rimasono d'acordo che assectandosi a cerchio quelli a chi toccha 9 andassino in mare. Onde uno christiano più astuto degli altri, ordinolli in modo che tucti i christiani rimasono et li giudei andorno in mare. Et a questo giuocho fare si accomoda più versi che inmediate qui seguiranno. Et di questi versi hai a pigliare le littere vocali. Et per questa littera A se ne ha a porre uno. Per la E se ne pone 2 perché è la seconda littera vocale. Et per lo I se ne pone 3. Et per lo O se ne pone 4. Et per lo U se ne pone 5. Et prima si comincia a christiani et seguitasi alternatim, cioè una volta giudei et una volta christiani et così seguendo. Et poi seguita sempre dove lassi et verratti facta et e' versi in su quali sono fundati
questi giuchi sono questi, cioè: | 78v

O tu che balli, balla bene chi t'apellerà.
Uno altro producente el medesimo:
Nove lapidanda et certi mantenea.
Uno altro producente el medesimo:
Populea ircha mater regina reserra.

Hora questi versi ti faranno sempre honore a fare questi giuochi, ponendo, comme è dicto, per le littere vocali che la prima è O et hai a porre 4 christiani, la seconda è U et tu hai a porre 5 giudei. Et così andrai ponendo et seguendo alternatim. Et farai el tuo quesito effecto et mai falla. Dicesi anchora dua versi latini che fanno la medesima conclusione, che dicano così:

---

13 assectandosi] F assittandoli | toccha] F toccava 14 astuto] K astatuto | ordinolli] F li ordinò 15 rimasono] F si salvarono 17 a porre] F a piglare 20 una volta giudei e una volta christiani] F una volta christiani e una volta giudei 20–21 et così seguendo et poi seguita] F et così seguita 21 in su quali] F in sulli quali 32 che dicano così] F quali dicano così

Quater, quinque, duo, unus, tres, unus et unus. Bis duo, ter, unus, duo, duobus, unus.

Pure cominciando a christiani et andando innanzi, cioè ponendo una volta christiani et una giudei, et poi cominciando a contare dove cominciasti a porre, etc.

Demostratio XVIIa generale regola a simili casi

79r Può avenire molti innumerabili casi com|me nel superiore a quelli che navichano et solcano el fluctuantissimo mare che mi contò una volta uno mio carissimo amico da Ragugia essersi trovato tre volte in grandissima fortuna di mare, che bisognato allegerire la nave di robba et di persone. Et una volta, giuchando alle buschette a nostro modo, li tocchò a essere gittato in mare. Unde lui prese uno pezo di legno, quello strectamente abracciato, spogliatosi nudo et invocato lo aiuto dello altissimo Dio et della gloriosissima Vergine Maria sempre vero refugio a miseri peccatori, si gettò nel profundissimo mare.

Et lui mi disse che immediate uscì di sé et secondo lui stette uno dì et dua notte fora di sé, poi si ritrovò nella Sicilia in casa di uno povero contadino che li disse ch'el mare lo haveva gittato fora. Et dua altra volte di poi erasi ritrovato giuchare al toccho che era scampato che non fu gittato in mare. Siché nelle fortune grande non si stima se non la vita, la robba et tutte l'altre cose si gettano
79v senza prezzo alcuno | et omni uno bada a quello che dice el patrone della nave che dice alle volte: "vedete, qui bisogna alleggierire la nave o noi periremo tucti.

Et meglio è che qualche uno perischa che tucti, siché fate fra voi quivi." Omni uno si racomanda a Dio et alle volte è bono sapere fare el conto tuo. Et perhò ti darò regola generale quando fussi uno che havessi uno conpagno a dua comme schade et havessi a fare al toccho. Et a chi toccha 7 o 8 o 9, secondo el pacto, paghi la tal cosa comme spesso interviene che si vole pagare una cena et che el tertio o el quarto ne vadi francho. Et chi resta nel cerchio ne vadi francho. Sempre farai così: cioè farai da te a te uno cerchio con la penna di quanti voi havete a essere fra tucti et comincia a contare donde tu vuoi et fa per 9 o per 8, comme ti pare et quando vuoi franchare 2 et tu chava tucti gli altri. Ovvero

9 Ragusa | fortuna di mare] F fortuna 12 spogliatosi] K spogliato 13 spogliatosi nudo et invocato lo aiuto] K et spogliato nudo et vogato con lo adiuto 13–14 sempre vero refugio] F sempre unicho rifugio 16 fora di sé] F fuori delli sentimenti 18–19 fortune grande non si stima] F nelle fortune non si stima 20 bada] K vada 21 tucti] F tucti quanti 27 et chi resta nel cerchio ne vadi francho] F cioè quelli chi restano nel cerchio ne vadino franchi 28 farai da te a te uno cerchio] F farai uno cerchio da te a te 29 et comincia a contare donde tu vuoi] F et cominciati dove tu vuoi 29–30 fa per 9 o per 8] F fa per 8 o per 9

segna che pognamo voi habbiate a essere 30 a una cena et 28 habbino a pagare uno grosso per uno, et dua ne habbi | a ire franchi, ingegnerati tu essere uno di 80r
quelli 2.

Farai da te uno cerchio di 30 "ooo" a questo modo. Et cominciati a tuo modo, solo segna con una croce dove cominci et poi anovera insino in 9 et quelli che toccha 9 et tu li punta nel mezo a questo modo: "ϴ". Et poi va segnando finché tu ne harai franchi 2 solamente, et noterai quelli due dove sieno et quanto stanno propinqui a colui che conta. Et ingegnati diporti quando hai a contare in uno di quelli luoghi, hora tu potresti dire che omni uno è astuto et malitioso.

Et io te respondo che gli è la verità, ma che noce el sapere questa cosa, non noce nulla el tentar la fortuna comme dice el corso: s'e' coglie colgha, se non havesti paura? Et così, quando havessi a essere 18, 20, 15 et 12, potrai aseptare el tuo abbacho et vedere in che luogo tu poi stare per essere francho et per 7, et per 9, o veramente tucti a dua numeri non noce.

Et sirà raro che se tu harai a mente e' luoghi franchi che non ti sappia o possa aseptare. Et qui si mostra quanto vaglia | essere bene experto in matematicha 80v
che giova omni cosa. Et per il contrario quanto sia innorme et dannosa el non essere bene experto ché hoggi dì la magiore villania che si possa dire a uno, si dice: "tu non sai quante dita tu hai nelle mani", ovvero "tu non sai quanto si facci 4 et 3", ovvero "non acozerresti 3 palloctole in uno baccino da barbieri" et molte altre simili cose, ché non si può dire peggio a uno, lassando da canto ladro et traditore et simili, ché omni uno non può essere philosopho, omni uno non può essere logicho, omni uno non può essere astrologho. Et quando uno non habbi li prenominate scientie, non li è vergogna, ma quando uno non è bono ragionieri li è grandissima vergogna. Et questa è una scientia che ciaschuno la può havere facilmente.

Demostratio 18 a ritrovare una moneta passata per via di cerchio o di qualche altra cosa

Havendo cominciato a narrare delli circuli, anchora voglio dare notitia di qualche piacevoleza che delli circuli si può chavare | ché per via di epsi potreno 81r
venire a molti mirabili effecti. Comme se tu fussi in qualche luogho fra la brighata et fussi a una tavola 15 o 20 persone, comme scade a piacere, et fussi donne et homini, et tu volessi fare qualche piacevoleza comme dire a qualche uno che "siché io te so dire a quella donna tu vuoi meglio di queste che sono

---

8 a colui] F a quello 9 tu potresti] F tu mi potresti 17 giova omni] F giova in omni 25–26 questa è una scientia che ciaschuno la può havere facilmente] F et questa scientia matemathica è di qualità che ciascuno la può facilmente imparare

qui", ovvero alle donne che "siché io vi so dire a chi voi volete meglio". Ma guarda non vi sia el marito, perché sempre in presentia sua la donna dirà volere meglio al marito et non sarà tenuto bello perché omni uno sa che la moglie dirà sempre volere bene al marito. Perhò starai in te, ovvero potrai fare uno cerchio di quatrini o grossi d'argento o ducati, et a tavola schacci o carte, come a te parrà, et dire "tochane una con lo animo". Et io ti saprò dire quale tu hai toccha senza alcuna interrogatione. Farai in questa forma: farai uno cerchio di quatrini o grossi, comme vuoi, che vi sia conio variato, perché l'amico possa discernere quale lui vole pigliare, perché se fussino tucti d'uno conio, con diffi-
81v cultà potrà ve|dere o eleggiere uno fra li altri, ovvero mescola grossi, quatrini et cratie, purché vi sia qualche differentia notabile. Et dirali a uno o a più, cioè a 3 o a 4 o a tucti, che omni uno pigli una moneta in core suo et tengola in sé. Potrai anchora aseptare le carte in cerchio et saranno specificate. Et comme l'amico ha preso, dirali che si scosti da quella che lui ha presa, quanto a llui pare et che anoveri, per che verso lui vole, o a mano ritta o mancha, non ti dà noia, et che anoveri insino a quella ha presa tacitamente da sé et solo ti dica da quale lui ha noverato che, verbi gratia, se lui si scosta 6 puncti da quella che lui prese et torni indrieto, anoverando insino a quella che lui prese, harà nel suo quore 6. Hora non voglio ti dica che sia scostato 6 puncti, perché non durerasti fatigha a ritrovalla che già lo sapresti, ma solo ti dica: "io ho cominciato a noverare da questa" et tocchila in modo che tu veggha, et "sono ito insino a quella che io
82r ho preso, contando a mano dextra". Hora | dirali che tengha in mente sua el numero che lui ha, cioè quanti puncti ha da quello che lui ti ha mostro alla sua pensata, et poi li dirai che in su quella medesima che lui cominciò a contare lui si ricominci et per regola generale sirà sempre a dua più che il numero delle monete, contando sopra el numero che lui ha in mente et andando al contrario che lui andò la prima volta che li contò. Comme se le carte o le monete fussino 20, sarà la sua pensata, contando al contrario sopra el numero che lui ha della prima numeratione, sarà alla 22 comme quivi vedi in figura da parte el cerchio el cui centro fia A et le figure tonde, cioè li "o" che tu vedi, sono monete di più sorte et sono per numero 20 monete. Et pognamo che l'amico habbi preso moneta nera et poi si tiri indrieto sino a quella che ha el puncto nel mezo, che sia tirato indrieto 9 puncti che cominciando a contare da quella che ha el puncto et andando a mano ritta verso la moneta pensata harà 9 puncti in mente. Et tu hai notitia che cominciò a noverare dalla puntata.

1 a chi voi volete meglio] F a quale uomo voi volete meglio 3–4 sempre volere bene] F volere bene 8 comme vuoi che vi sia] F a tuo modo che vi sia 12 in core suo] F in suo cuore 15 o mancha] F o a mano mancha 27 le monete] K la monete 29 numeratione sarà alla 22] F numeratione alla 22 poste 30 cioè li "o" che tu vedi] F che tu vedi, cioè li "o" 34 pesata] F pensata

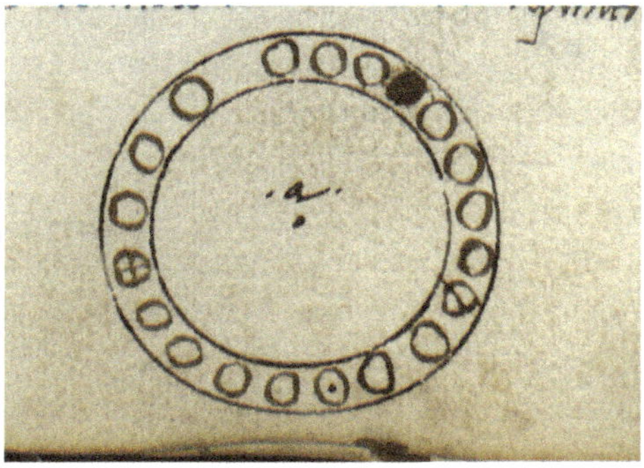

FIGURE 4
Drawing at the bottom of folio 82^{r}

Ho|ra tu, fermandoti quivi, dirai che anoveri per il contrario che lui prima 82^{v} anoverò et sarà alle 22 la sua pensata moneta che se tu anoverarai in su la puntata allo in qua, cioè a mano sinistra anoverando sopra 9 che lui ha alle mani alle 22, sarà la moneta nera che lui ha presa. Hora per più mirabilità potrai dare libertà allo amicho, permectendoli fare venire la sua moneta a quanti puncti lui vole. Et mectiamo che una altra volta habbi preso quella moneta che ha la + nel mezo. Hora dirai che si scosti a suo libito et cominci in su una comme di sopra, et anoveri insino a quella che lui ha presa. Et per che verso vole, che a te non porta. Hora diciamo che lui si sia fermo in su quella del puncto nel mezo et vadi a mano dextra che, numerando, harà nel suo quore i 6. Hora tu li domanderai a quanti puncti lui la vole che li farai venire, notandoli che lui dica magiore numero ch'el numero delle monete et dua più, perché dicendo minore numero non potre|sti fare lo intento tuo et suo. Ma per non li fare pigliare malitia, starai 83^{r} atento al numero che lui dice che forse per sé medesimo lo passerà, ma quando non lo passassi dirali tu allora che vuoli magiore numero che passi, verbi gratia, 25. Et se lui dirà: "io la voglio alle 25 puncto", et tu tacite con l'occhio a ciò niuno possa tua facilità et regola conphrendere, comincia in su la sua che lui prese, cioè in sulla moneta punctata, et dirai 22. Et poi andrai nel verso che contò lui, tornando indrieto, dicendo 23. Et poi una altra più, la pure nel verso che contò lui et dirai: 24. Et poi andrai una altra pure in quello medesimo verso et dirai: 25 et sarai tornato indrieto 3 puncti. Et omni volta che tu vai al verso che contò lui la prima volta, tu vieni a tornare indrieto. Et perhò, se bene porrai mente volendola alle 25, ti bisognerà cominciare in su quella moneta che sta così: "ϴ" nel cerchio. Et andando al contrario che lui ha anoverato, cioè se lui è ito a mano dextra et tu va a sinistra et se lui è ito a sinistra et tu va a dextra, troverai la

---

4 sarà la moneta nera che lui ha presa] F sarà la sua moneta pensata, cioè la moneta nera 6–7 ha la + nel mezo] F ha la + 12 delle monete] K della moneta 13 fare pigliare] F fare piglia 16 alle 25] F alle 23

83v mon|eta per lui pensata che se lui conta sopra quella et sopra el suo 16 che epso ha a 25, farà quella della croce che sta così "O" comme lui pensò. Et se dicessi: "io lo voglio alle 60", in questo caso dico che tu cominci in su quella che cominciò lui, et dica 22. Et andando, anoverarai per il verso che anoverò lui, tanto girarai el cerchio che lui, tu arriverai all 60 et dirali che anoveri al contrario che hai anoverato tu et verratti facto. Ma non li dire: "anovera el contrario di me", ma mostrali con mano, per che verso lui ha a noverare. Et se bene lui la volessi alle 100 o alle 1000, tenendo la data regola potrai fare venire a che numero vuoli. Potrai anchora fare venire a mancho che 22, quando el compagno lo richiedessi, alle 18, 20 o 16, tucto potrai per farlo più mirabile, ma in questo caso bisogna che tu ti ponga leggie che lui non passi in el suo tirarsi indrieto numero di 8, 10 o 12, 84r al più. Et quando lui havessi preso una moneta et tiratosi indrieto 10 puncti | et tu volessi farla venire alle 18 et tu in su la moneta che lui pigla ti ferma et va innanzi 4 puncti, dicendo in su quella 22, in su l'altra 21 et in su l'altra 20 et poi 19 et poi 18, et dirali che anoverando sopra quella alle 18, sarà la moneta pensata. Et nota che quando tu ti fai innanzi che tu vuoi scemare comme qui che tu vuoi che la vengha alle 18, hai a noverare per il contrario che anoverò lui. Et se la volessi alle 16, saresti ito innanzi dua puncti più et sarebbe stata alle 16. Più oltre non è da procedere se già tu non pattegiassi che lui non si potessi tirare indrieto più che 4 o al più 6. Et allora potresti farla venire alle 10 o 12 a suo libito. Starai dunque vigilante a quello bisogna et harai honore.

Demostratio 19a a trovare uno numero pensato per via di cerchio altrimenti

Fassi anchora una altra mirabil forza per via di cerchio: sapere trovare uno pensato numero per lo amico. Et fassi in questa forma: a te bisogna aseptare uno 84v cerchio di numeri co|perti che niuno altro che te sappi comme stanno et sia di che quantità si vole. Potrai torre polize segnatovi su in su la prima 1 et in su la seconda 2, in su la tertia 3, etc. et potrai procedere in infinito. Ma a questo ti potranno servire benissimo li germini che sono signati per numero, ciaschuno potrali aseptare coperti prima uno et poi 2 et poi 3, andando a che mano tu vuoi che non fa caso, pure che a te sia noto a che mano tu li hai aseptati, ma per più tua facilità asepterali a mano destra che ti sarà mancho laboriosa alla tua fantasia. Hora diciamo che tu habbi aseptato 20 poste in tucto coperte et a te solo note comme stanno. Et sempre terrai l'occhio dove stia la unità, perché quella è tua guida. Et se le brighate volessino vedere comme stanno le poste coperte, questo per te si neghi, perché non saria poi tenuto el giuocho bello. Et perhò, quando tu cominci a fare e' giuochi dirai alta voce: "se voi volete che io facci

---

3 dico] F di 13 alle 18] F alle 12 18 puncti più] F puncti 22 altrimenti] K alitrimente 35 tenuto el giuocho bello] F tenuto bello

qualche gentileza, state quieti et a vedere et che quivi se taccia, perché | così se [85r] fa a bono conpagno a non guastare li giuochi l'uno al'altro, siché omni uno stia quieto, fermo." Et a vedere hora comme tu hai aseptato le tue 20 poste, dirai: "omni uno di voi pigli uno numero a suo modo quale non passi", specificando el numero delle tue poste, le quali qui sono 20. Onde dirai niuno non passi venti. Quando tu aconciassi e' germini tucti, saria 40. Et potresti dire piglassino sino in 40, ma per hora fermiamoci in sul 20. Et tu sempre pigla per tua guida uno numero che sia più 2 che la tua posta che in questo hai 20 poste. Dunque piglia per guida 22 et se fussino 40 poste, haresti a pigliare 42 et così in infinito. Dunque in questa hai per guida 22 et questo 22 lo ferma in su la posta che vi è sotto la unità. Et dirai chi ha preso si facci innanzi. Et dirai che annoveri in su la unità sopra el numero che lui ha preso. Et vadi a mano sinistra, cioè verso la posta magiore, cioè verso el 20 et che alle 22 sarà sotto la posta el numero che lui prese, ché se lui | harà preso 18 che dica in su la posta della unità 19 et seguiti [85v] in quel verso sino in 22. Et alle 22 si fermi et scopra quella posta. Et quella sarà 18, cioè el numero che lui prese. Et se fusssino 100 che havessino preso, a tucti quanti riuscirà che tucti troveranno el loro numero, ma perché la brighata non si accorgha che sempre tu vada in su la unità alle 22, et potrebbero facilmente conphrendere el tuo giuocho. Et perhò, tu volendo mutare, andrai in su l'altra a lato ala unità, cioè a mano dextra in sul 2, et dirai 23. Et poi in sul 3, et dirai 24, et poi 25, et 26, 27, etc. Potrai comme nella superiore andare alle 100, alle 1000, in infinito. Et perhò, tu muterai cibo: hora alle 22, hora alle 23, et quando alle 30, et talvolta alle 29. Et anchora potrai fare el giocho più mirabile, dicendo allo amico, comme ha preso el numero, che ti dica alle quante poste lui vole che tu la ffacci venire ché ti obblighi a fare venire a quante lui vole et poi li dirai procedendo comme di sopra è dicto, dicendo: "annovera qui | su" et verratti [86r] comme se lui la chiedessi alle 24. Et tu ti scosterai puncti dua dalla tua unità. Et se lui la chiede alle 30, et tu ti scosterai 8 puncti dalla unità, et se alle 40, et tu ti scosterai 18 puncti dalla unità, et così in infinito. Se lui la volesse alle 1000, sempre la farai venire. Et nota che nel chiedere a che numero lui vole che tu li facci venire, lui non chedessi mancho che alle 22, che ti potrebbe fallire la regola. Quando el numero che lui ha preso fussi 20 et lui la chiedessi alle 18, sarebbe falsa la regola. Et perhò dirali che volendola alle mancho che alle 22 che non passino a pigliare numero 8 o meno, et che la farai venire alle 20 o alle 18, o alle 16, comme lui vole, comme nella superiore dimostratione fu decto. Et quando si obligassi a non passare 4 di numero, cioè piglerà uno numero da 4 in giù, potresti fare venire alle 12 o 10, comme volessi, tenendo el modo comme

---

15 et scopra] F et sopra 25 che ti obblighi] F che ti obli 32 la chiedessi alle 18] F la chiedessi alle 8

nella superiore, cioè quando lui havessi preso 8 et volessi che tu li facessi venire el suo numero alle 18. In questo caso tu andrai a mano sinistra 4 poste innanzi,
86v dicendo in su la unità | 22. Et poi in su la posta del 20/21. Et poi in su la posta del 19/20. Et poi in su la posta del 18/19. Et in la posta del 17/18. Et dirai che, contando sopra el numero che lui ha alle mani, che alle 18 verrà la sua chiesta, cioè el numero che lui prese, ma tucto farai in dextro modo ché l'amico non si accorgha della tua astutia o regola.

Demostratio xxa per la sopradecta regola a fare uno giuoco con le carti

Fassi per questa forza uno giuocho bello con le carti che si mette uno asso, uno 2 et uno 3 et 4, insino 19. Quando le carte hanno 10, si mette sino in 10 et poi fante, cavallo et re. Alle carti che hanno 9 si pone nome, al fante 10 et al cavallo 11 et al re 12. Et alle carte che hanno 10 si pone nome, al fante 11 et al cavallo 12 et al re 13. Et alle carte che hanno reina si pone 13 et al re 14. Et questa valuta farai nota alle brighate a ciò lo sappino. Et tu, non ponendo tuo facto, tenendo
87r le carti in mano, ne caverai una di | omni numero, cioè, comme è dicto, uno asso, uno dua, uno tre et così de omni ragione figura, cioè uno fante, uno cavallo, una reina essendovi, et uno re. Et poi asepta tucte dicte carte ordinate per cerchio in modo vengha tondo apuncto, a più confusione delle brighate. Et non dicendo quello vuoi fare, staranno suspesi. Et tu, aseptato che harai el tuo cerchio, bene tenendo sempre firma la tua unità dove sia che senza havere omni volta a guardare la carta tu possa fare el bisogno, et poi piglierai le carte et fra tucta la brighata le dividerai, dandone a chi 4, a chi 6, etc. Et poi ti comincia a uno dicendo: "piglia una delle tue carte, quale vuoi, che non fa caso, et poni mente che sia quella tale carta. Et dimmi a quanti puncti tu vuoi che io la faccia porre", dicendo che passino el tuo numero, cioè el numero che è tua guida, a ciò meglio et con più facilità possa fare el giuocho che pognamo dica
87v alle 18 poste volerla. Et tu farai così: mettiamo che | le carte sieno comme nella nostra ciptà s'usano, che di puncti sono insino a 9 et hanno 3 figure: re, cavallo, fante, che in tucto sono 12 carte per giocho et in tucto sono 48 carte. Hora tu harai preso 12 carte. Et nota che quando tu le pigli, che tu non pigli tucti bastoni o tucte spade, perché la brighata si poterebbe acorgere della cosa. Et pigliando d'omni giocho non potrannno tucto vedere et maxime, perché tu le stribuisci fra più persone. Dunque, tu harai aseptato el tuo cerchio di 12 poste, et tu sai per le superiori dimostrationi che tu hai sempre per generale regola a porre in sul numero delle poste dua. Dunque, harai 14 per tua guida et questo 14 fia fundato in su la unità. Dunque, dirai colui che ha preso la carta che numerando sopra la carta, da te mostrali—che li mostrerai la unità—et contando sopra el numero

2 el suo numero alle 18] F el suo numero alle 12

della carta vadi sino alle 14 et quivi si fermi et posi la carta che lui ha in mano senza guardarla, overo se lui ti harà dicto | volerla alle 18. Et tu dirai in su la 88r unità 14 et poi, procedendo innanzi a mano dextra, andrai dicendo: 15, 16, 17, 18. Et dirali qui su contando a mano sinistra sopra el numero della carta, poserala dove ti guida 18. Et se la carta è 5, la poserà per forza in sul 5 che quivi hai posto. Et se 6, la poserai in sul 6. Et se 7 la poserai in sul 7. Et se fante, in sul fante, etc. Et così dirai a tucti per fino che tu harai facto posare tucte le carte. Et poi tu dextramente le racogli a 4 a 4, ponendole ordinariamente et anche con dextro modo potrai mescolarle o fare vista, levandone 10 o 12 di sotto et mectendole di sopra. Et quando hai facto quello vuoi, et tu le scopri alli astanti che non sanno certo anchora quello ti voglia fare. Et mostrali che tu hai acozato tucti li assi insiemi et tucti li 2 et tucti li 3 ordinariamente, ché in verità sarà tenuto giocho bello apresso le brighate. Et potrai fare molti altri giuochi che con la tua fantasia li funde|rai in questa legiadra regola. 88v

Demostratio XXIa a trovare una carta cavata a vedere le carte due volte solamente

Poi ch'io sono entrato nelle carte, voglio mostrare in questa mia dimostratione uno modo da potere trovare una carta che fussi stata tracta di tucte la carte, che sia per forza et virtù di numero. Terrai questo ordine et modo: porrai tucte le carte in tavola et dirai: "cavate una carta et io le voglio vedere a una a una tucte due volte et saprò dire che carta voi havete cavata". Et comme loro hanno cavato la carta et tu piglierai le carte et porrai mente. Nota che la prima volta che tu le vedi tu hai a porre mente che numero è, o se è 8, o 2, o 9, o figura. Et la seconda volta che tu le vedi hai a porre mente che giocho è quella tale carta. Hora per vedere la prima volta che numero ha in sé la carta che mancha, terrai questo modo: prima bisogna che sappi le carte di che qualità sono: se hanno 8, o 9, o 10, ché alcune carte non hanno più che 8 | puncti et 3 figure, cioè re, 89r cavallo, fante, et alcune hanno 9 puncti et 3 figure, et alcune hanno 10 puncti et 3 figure et altre hanno 4 figure, re, reina, cavallo et fante. Hora comme nella nostra florida ciptà si usa, diremo sieno carte di 9 et 3 figure, che in tucto sono 48 carte. Hora bisogna che al fante pongha nome 10, el cavallo 11, et al re 12. Et poi somma da uno sino in 12 quanto fanno, ché numerato da uno sino in 12 tucti i numeri che vi sono inclusive, farà 78. Dunque, per omni giocho tu hai 78 et sono 4 giochi. Et perhò dirai 4 via 78, fa 312. Et hai tucta la somma delle carte che in tucto sono 312, quella dell'octo, solamente sono tucte per numero 264, perché sono per giuocho 66 puncti. Hora attendiamo alla nostra. Tu sai che

14 legiadra regola] F regola legiadra. 35 dell'octo] F dell'octo, del 8

tucta la somma è 312. Hora tu piglierai le carte et bisogna che habbi el cervello
techo et somma tucte le carte da te che tu hai insiemi, dicendo: 4 et 7 fa 11, et
89v 10 fa 21, et 5 fa 26, et 12 fa | 38, etc. Et guarda di non errare mai, ché se tu errassi
di uno puncto solo, non faresti el bisogno. Et questa somma la farai tacita da
te che dalli circunstanti non sia sentito ché conphrenderebbono la tua regola.
Et perhò, da te tacite sommerai tucti i puncti delle carti che ti trovi in mano. Et
quando l'hai sommate tucte, porrai mente quanto ti mancha a fornire la somma
di 312 et quanto ti mancha tanti saranno e' puncti che la carta che mancha, et
se mancherà 10, sarà fante, se 11, sarà cavallo, se 12, re, etc. Hora, se tu sapessi
quale re manchassi, cioè di che giuocho si fussi, tu haresti el quesito, ma perché
tu non sai che giuocho si sia, perhò bisogna nella seconda veduta vedere quale
giocho mancha. Et questa seconda volta potrai correre le carte presto et fare
quasi vista. Non ti curare di vederle, ovvero quando al 1/3 o al 1/2 o al 2/3 di
decte carte tu hai veduto el tuo intento, ché spesso interviene che a 1/2 le carte
90r tu hai veduti 3 re, o 3 fanti, o 3 cavalli o 3 9 che | tu vuoi vedere di 10, non voglio
vedere più. Questo mi basta et parrà più mirabile et tu dirai: "havete cavato
la tale carta specificandola". Et bisogna che a fare questo tale giuocho tu sia
buono abachista et buono praticho nel racorre, perché quanto più presto corri
le carte, tanto più mirabile fia tenuto. Et guarda non errare, perché faresti ride-
re le brighate che uno puncto che tu erri, ti guasterà el disegno. Ma tenendo el
cervello a botegha non potrai errare et verratti facto infallanter et sarà tenuto
giocho bello et di fantasia.

Demostratio XXIIa a trovare una carta tracta di tucte le carte a vederla una sola volta

Possi anchora a vederle una sola volta havere notitia di una carta che fussi stata
tracta di tucte. Et benché sia caso difficile a farlo, pure lo voglia in questa nostra
opera inducere. Et il bono praticho abachista lo farà benissimo et con pocha
90v faticha. A te bisogna prima le carte tucte a nome battezare et | che tucte hab-
bino uno numero di per sé. Et prima porrai nome a bastoni, andando, per non
errare, drieto all'ordine della a, b, c, cioè prima bastoni, poi coppe, poi danari, et
poi spade. Et a bastoni, cioè al'uno bastone, porrai nome uno et 2 et 3, di mano
in mano, fine in 12. Et poi al'una coppa porrai nome 13, et alle 2 coppe 14, et alle
3 15, et alle 4 16, et di mano in mano, sino al re di coppe, harà nome 24. Et poi
porrai nome al'uno danaio 25, et a 2 26, et a 3 27, et a 4 28, di mano in mano per
insino al re di danari che harà nome 36. Et poi porrai nome al'una spada 37, et
alle 2 38, et alle 3 39, et così andarai discurrendo, per insino al re di spade porrai

2–3 et 10 fa 21, et 5 fa 26] F et 10 fa 11, et 5 fa 16 15 o 3 fanti, o 3 cavalli] K o 3 fanti

nome 48. Et havendo bene questi numeri aseptati nella mente tua, che bene li habbi in praticha, potrai fare questo mirabile giocho. Hora bisogna che tu racolgha tucti li numeri che inclusive sono da uno insino in 48 et che per la regola
or|dinaria lo farai presto ché legendo questo mio libro anchora a questo darò 91r
regola. Hora, racolti e' numeri da uno insino in 48 inclusive, harai una somma di 1176 per tucta la somma. Hora, comme l'amico harà cavato la carta che lui vuole et tu pigla le carte in mano et asommerai tucte le carte, ponendo sopra l'uno al'altro i numeri, sino che harai acozate tucte le somme et vederai quanto ti getteranno tucte et vedrai quanto ti mancha che se ti mancha da 12 a uno, sarà bastoni. Et se ti mancherà da 24 a 13, sarà coppe. Et se ti mancherà da 36 a 25, sarà danari, et se da 37 a 48 sarà spade, che havendo tu bene e' numeri a mente, non ti poterà fallire. Solo bisogna che tu usi diligentia nello asommare le carte ché, comme ho dicto, sarà caso difficile a chi non è bono praticho. Et anche a uno bono praticho sarà uno poco fantastico, perché hai dua difficultà: la prima a sapere è 7 danari, che numero lui ha, che e' 7 danari ha di numero 31, et poi
acozarlo con l'altra | somma. Ma tu, per più facilità, potrai cavare e' 9, et reste- 91v
ranno carte di 8. Et harai mancho somma ché in tucto harai 44 carte. Et harà el re di bastoni nome 11, et l'una coppa 12. Et così di mano in mano l'ultimo, che sarà el re di spade, harà nome 44. Et harai per tucta quanta la somma di tucte le carte 990, che a te sarà più facilità, siché piglerai quello modo che a te parrà più acomodato a tua fantasia. Et omni uno sarà bello giocho.

Demostratio 23a a trovare 2 puncti di dua dadi gittati a sorte

Anchora per la forza, da te, del numero, potrai pervenire alla notitia di dua puncti di dua dadi tracti a sorte, ovvero a sorte presi comme si sia. Daremo vera regola acomodata a questo caso. Et dirai a uno che pigli 2 dadi et che con epsi tragha 2 punti a te incogniti. Et che tu ti vanti per certa multiplicatione
ritrovarli et comme lui ha tracto, dirai che pigli uno di decti puncti et dica|ti 92r
se doppia el minore o el magiore puncto.* Et poi sopra el doppiato del puncto ponga 5. Et poi el tucto multiplichi per 5. Et sopra el producto di questa multiplicatione pongha el puncto del secondo dado et poi ti dica la somma che la dirà. Et di questa somma caverai 25 et quello resta, le decina, sarà el puncto del dado di quello che radoppiò. Et el numero sarà el puncto del secondo dado. Potresti fare cavare a llui quello 25, ma se lui havesse ingenio conphrenderebbe la tua regola. Ma bene puoi farli cavare 6, o 8, o 11 et poi dicati el resto. Et tu caverai insino in 25 et harai el quesito. Ecchoti lo exemplo: sia che il compagno

10 et se ti mancherà] F et se ti mancha mancherà 24 presi] K prese

* a margine una nota: *vacant*

habbi tracto 6 et 4, doppi uno, et sia 6, fa 12, postovi su 5, farà 17, et multipli-
cato per 5, fa 85. Postovi su quello altro puncto, cioè 4, fa 89. Et pognamo che
tu li dica: "della sua somma cavi 11", resterà 78. Hora dirai ti dica la somma che
lui ha alle mani, dirà: 78. Et tu caverai el resto sino in 25 et 14, che cavato 14 di
92v 78, resterà 64. Eccho che la d|ecina che sono 6 decine saranno el numero che
facesti doppiare. Et il numero sarà el secondo puncto. Et se lui havessi doppia-
to el minore fa 8, postovi su 5 fa 13, multiplicato per 5 fa 65, postavi su 6, cioè
l'altro puncto, fa 71, tractonne 25, resterà 46. Et le decine sono el puncto che
facesti doppiare. Et se bene havessi tracto buffa, cioè 44 et 66, per la medesima
regola servirà che doppia uno, cioè 33 fa 6, ponvi 5 fa 11, multiplicha per 5 fa 55
et postovi su 3 fa 58, cavatone 25, resterà 33. Eccho che la regola servirà in tucti
e' modi. Et per questa regola potrai anchora dire a uno che divida 10 in 2 mani,
et saprai dire quanto habbi per mano. Et anchora potrebbe dividere 12 et 16, et
per insino in 18, ma dividendo in modo che non vengha numero per mano, che
passi 9, ché non ti servirebbe questa regola, comme a dire 9 et 9, ovvero 9 et 8, et
7 et 8, etc. Et potrai fare li altri con la demostratione sopra decta a trovare uno
93r numero diviso in 2 parte, etc. Ma questa regola serve a dua numeri che non |
passino 9 per uno et è bella et bona, comme vedi, etc.

Demostratio XXIVa a trovare 3 puncti di tre dadi

Et se fussi 3 dadi tracti a sorte, opure presi aposta, quomodocumque sit, et
tu volessi e' loro puncti investigare, punctualmente anchora daremo uno altro
breve modo a ciò fare acomodato. Et fia questo: comme lui ha tracto e' dadi,
dirali che pigli uno delli puncti qual vuole et che lo adoppi. Et poi in sul doppio
ponga 5. Et poi tucto multiplichi per 5 et di poi pongha su el puncto del secondo
dado qual vuole. Et tucto multiplichi per 10. Apresso vi pongha su el puncto del
tertio dado et dicati tucta la somma. Et tu, per generale regola, cava sempre 250
di quello resta et le centinara saranno el puncto del primo dado che fu doppiato
et le decine saranno el puncto del secondo dado et el numero sarà el puncto del
terzo. Pognamo che lui habbi tracto o preso 6, 4 et 2, et poi habbi doppiato 2,
fa 4, et in questo pongha 5, fa 9 et multiplichi questo per 5, fa 45. Et pongavi su
el secondo puncto sia 4, fa 49. Et questo multiplichi per 10, fa 490. In su questo
93v pongavi el pun|cto del tertio dado che è 6, fa 496. Hora dico che tu ne cavi 250,
hora potrai fare cavare a llui 125 o 100 o 225 o comme più copertamente tu puoi,
cava el resto che a dire a llui "cavane 250", facilmente potria conphrendere la tua
regola. Et perhò, dirali ne cavi una parte. Et se anchora tu ti fai dire el numero,

5 saranno el numero] F farà el numero 20 aposta] K apostatamente 26 per generale regola] F per regola generale

absolutamente può anchor suspicare. Et perhò poi li mecterai el cervello a partito, dicendo ne cavi 125 o 130, etc. Ma diciamo ne facci cavare 125, che cavato 125 da 496 resterà 371. Et farai ti dica questo 371 a te, et tu ne caverai altri 125, che resterà 246. Et harane cavato 250, et questo 264 saranno i puncti del dado, cioè li dua centinara denotano el primo puncto che facesti doppiare et le 4 decine denotano el secondo puncto del dado, et li 6 puncti ti denotano el tertio puncto. Et così dirai che traessi 2, 4, 6. Et potrai per questa regola ritrovare 3 numeri divisi fra 3 persone, purché non passino li decti numeri 9 per uno, perché la decta regola non servirà. Et così potrai | ritrovare 3 carte, purché non sieno fi- 94^{r} gure ché potrai porre 3 carte in tavola a te note et dire: "3 di voi le piglino et io voglio investigare che tanta voi havete preso ché fia bella forza di numero".

Demostratio xxva a trovare 3 dadi per altra regola

Per una altra regola et modo quasi simile alla superiore, potrai ritrovare 3 puncti di 3 dadi che fussino stati tracti o gittati per lo amico tuo in questo modo: cioè dirai che uno di decti puncti adoppi et poi vi pongha 5 et tucto multiplichi per 5 et poi vi pongha su 10 et poi el punto del'altro dado, el tucto multiplichi per 10 apresso el puncto del'altro dado. Et poi dirai che della somma traggha 350, el restante, comme nella superiore, sarà le centinara el numero del primo et le decine del secondo et el numero del tertio dado. Comme se lo amico havessi tracto 5, 4, 3 dirai adoppi uno sia 4 fa 8, postovi su 5 fa 13, multiplicha per 5 fa 65, pongavi su 10 fa 75, pongavi su l'altro dado sia 3, fa 78, multiplichi per 10 fa 780, pongavi su 5, cioè l'altro dado, fa 785. Cava di questo 350 | o, comme nella supe- 94^{v} riore, farai cavare a llui in dextro modo la metà, ovvero el terzo in modo che mancho che sia possibile possa conphrendere la tua regola. Che in questo caso harai 785, cavane 350, resterà 435. Eccho che li 4 centinaia ti rapresentano e' 4 puncto che faresti doppiare. Et le 3 dicine ti rapresentano el secondo numero, el 5 ti significa el tertio dado. Et quando bene el numero fussi tucti 3, tucti 4, tucti 5, ti riuscirà comme se fussino tucti 2, doppia uno fa 4, poni 5 fa 9, multiplicha per 5 fa 45, poni 10 fa 55, poni uno altro puncto fa 57, multiplicha per 10 fa 570, poni l'altro dua fa 572, cava 350 resterà 222. Et così dirai che traessi 222. Et mai ti fallirà. Con questa et con la superiore potrai fare vari et innumerabili giuochi fondati in su queste regole, pure che li numeri si pigliano non passino 9 per numero et con carte et con dadi et con chiose et quarteruoli, a tuo modo. Et tieni bene a mente questa regola, quando el numero passassi 9 hai le demostrationi dicte nel principio di questa seconda | parte ché ti insegnamo fare giochi 95^{r} mirabili, comme puoi vedere. Hora andiamo innanze.

---

17 traggha] F traghino 23–24 terzo in modo che] F terzo che 34 passassi 9] F passi 9

Demostratio XXVIa a fare omni puncto con 2 o 3 o 4 dadi a chiusi occhi

Farai anchora uno mirabile giuoco nel conspecto delle brighate, che fia questo. Dirai alta voce: "io voglio fare omni puncto che voi volete con li dadi a chiusi occhi, cioè non voglio niente vedere", il che certamente sarà di admiratione a chi non sa sua facilità. Faralo in questo modo: dirai ti dichino quanti puncti vogliano che tu farai o 20 o 30 o 50, o quanto magiore sarà più mirando et più facile a te. Et perhò dirai che chiegghino gran puncto che li piccoli sono facili. Hora diciamo che loro ti dichino 24. In questo caso tu hai a notare comme per altra ti ho decto, che omni dado fra sotto et sopra è 7 puncti. Et essendo tu di questo certo, potrai fare questo giocho in questa forma. Parti el numero che tu o loro vogliono per 7 et verranne 3. Dunque 3 dadi bisogna trarre. Et se tu non
95v ne havessi se non | uno, trallo 3 volte. Hora, partito 24 per 7 ne viene 3 et avanza 3. Hora certa gratia tu dirai alli astanti pigliare questo puncto a occhi aperti, mostrandoli 3 ché questo 3 ti saria difficile farlo a chiusi occhi. Et poi pigla li dadi preso che loro hanno el tre, et trarali con dextreza et dirai che piglino el puncto che gettano li 3 dadi. Et poi acortamente piglerai li dadi in modo non paia tuo facto. Et tucti 3 li volterai sotto sopra havendoli coperti con la mano. Et poi li scopri. Et dirai: "piglate questo altro puncto et con el 3 ch'io vi detti, fa 24 apuncto". Et mai ti fallirà. Et se non havessi se non uno dado, traralo 3 volte, et omni volta acortamente volteralo sotto sopra, dando li puncti di sotto et di sopra, ma più bello parrà a fare con 3 dadi. Et quando volessimo 50 puncti, potrai fare con 4 dadi ché in 50 vi sia 7 dadi et avanza uno. Onde tu piglia uno, dandolo loro a occhi aperti. Et poi thrarai 4 dadi una volta. Et poi con gratia
96r li volta sotto sopra, traendone 3 et mede|simamente li volterari sotto sopra et harai 49, et uno che li desti prima fa 50 et harai facto 50 a puncto. Et se bene volessino 100, 1000, potrai fare con questa regola, sempre partito el numero che loro vogliano per 7 et quello avanza dallo loro al primo ché non pare sconveniente dare el primo numero a occhi aperti. Ma se loro chiedessino numero che partito per 7 non avanzassi nulla, in questo caso non bisognerebbe che tu li desse numero con li occhi aperti. Et perhò, quando ti dicessino volere 35, 42 o 70 et simili, allora parrà più stupendo. Et nota che simili giuochi non si vogliano fare più che una volta, perché facilmente sono conpresi dalli acutissimi ingeni. Et potrai fare molti mirabili effecti derivanti da decta regola.

Demostratio XXVIIa a trovare uno anello fra più persone a quale mano et a quale dito et a qual nodo

Fassi una regola a ritrovare uno anello fra più persone fussi stato nascosto, et
96v sapere dire in qual mano, in qual dito et in qual | nodo lui l'ha messo. Et fassi in questo modo: per la regola delli 3 dadi dirai a tucte le persone che si aseptano a fila, et poi dirai a uno che sia alquanto pratico nel multiplicare, cominci a sé

et anoveri insino a quello ha l'anello. Et dirai che questo numero lo doppi. Et poi vi pongha su 5 et poi lo multiplichi per 5. Et in su la somma ponga 10. Et poi vi pongha su el numero delle dita a quanti lui lo ha, cominciando a contare dal dito minore della mano sinistra, ch'el dito grosso della mano sinistra harà nome 5 et el dito grosso della mano dextra harà nome 6. Et così di mano in mano, l'ultimo dito della mano dextra harà nome 10.

Hora, posto che lui harà el numero delle dita et tu caverai o farai cavare di tucta la somma 350. Et quello resta: le centinara sarà el numero della persona che ha l'anello, le decine saranno a quanti dita lui l'ha. Et per consequenza, havendo a quale dito, harai a quale mano, el numero ti denoterà a quanti nodi. Et pognamo che l'havessi a 5 homo et a 6 dita et a 2 nodi.

Dirai a colui che adoppi el numero delli homini, harà 10, postovi su 5 farà 15, 97r
multiplichato per 5 farà 75, et a questo giungha 10, farà 85. Et poi vi ponga su el numero delli dita che vi porrà su 6, farà 91. Et questo dirai, multiplichi per 10, farà 910. Et a questo agiunghi el numero delli nodi, che sono 2, farà 912. Hora farai thrare overo thrarai tu de 10 che ti farà la somma, overo farai parte thrare et parte thrarai tu, in tucto si ha thrare 350, che tracto 350 di 912, resterà 562. Et li centinara, comme è ditto, ti notifichano el numero delli homini et le dicine el numero delle dita et per consequenter la mano el numero ti mostrerà li nodi, comme qui ti mostra 5, 6, 2. Dirai che l'ha el quinto homo al dito grosso della mano ritta al secondo nodo. Ma nota che quando lui lo havessi decimo dito, cioè nel dito mignolo della mano dritta o dextra che sarebbe el decimo dito, in questo caso le decine non potrebbono mostrare le dite et le centinara non potrebbono mostrare la verità, perché sarà uno centinaro più. Et | perhò, nota 97v
che quando el numero che l'amico dice, cavatone 350, non vi resta dicine nessune, ma solamente centinara et numeri comme 703 o 402 et simili, in questo caso piglierai li centinaia, et caverane 1 et el resto sarà el numero dello homo che ha lo anello. Et quello centinaio che tu cavi sarà per li 10 dita alle quali lui è, comme se el quinto homo havessi lo anello, al decimo dito al primo nodo, et anchora bisogna tu te intenda con colui che anovera, donde si fa anoverare i nodi o dalla puncta del dito o dal ceppo. Et perhò, dirali si faccia dal ceppo del dito et vengha verso la puncta a numerare. Hora, per tornare al giuocho nostro, dirai che adoppi el numero delli homini, cioè 5 farà 10, ponga su 5 farà 15, multiplicha per 5 farà 75, et ponga 10 fa 85, et ponga el numero delli dita, cioè 10, fa 95. Et questo multiplichi per 10, fa 950. Et a questo ponga el numero de' nodi, che è 1, fa 951. Hora di questo caverai 350, resterà 601. Et li centinaia denotano 6 homini et fuorno 5, ma cavandone uno resterà 5. Et quello uno centinaio te

---

16 de 10] K detto 29 al decimo dito al primo nodo] F al primo nodo

98r deno|ta 10 decine per li x dita, et uno ti denota el primo nodo, etc. Et mai ti fallirà questa regola.

Demostratio 28a a trovare el medesimo anello per altra regola

Potrai per una altra legiadra regola venire alla sopradecta cognitione et investigatione. Et fassi in questo modo: cioè dirai comme nella superiore a uno che sappi alquanto multiplichare che cominci da sé a contare, et vadi insino a quello che ha l'anello, et anoveri inclusive, cioè anoveri sé et quello che ha l'anello. Et poi dirai che adoppi el numero, et a questo agiungha 5, et multiplichi tucto per 5, et pongavi su 10. Et poi dirai se lui l'ha nella mano dextra, vi pongha su dua, et se lui l'ha nella mano sinistra vi pongha su uno. Et dirai tucto multiplichi per 10. Et a questo pongha el numero delle dita, cominciando a contare omni mano al dito grosso o al mignolo non porta, ma a te sia noto donde lui comincia a contare. Et se l'ha nel minimo della mano dextra, dirai 5, et così
98v l'altra mano di per sé ché in qu|esto caso non si mescolano le dita insiemi, ché omni mano conta sua dita di per sé. Et perhò, diciamo che pongha el numero delle dita, et poi tucto multiplichi per 10. Et poi pongha el numero delli nodi che cominci a contare dal ceppo, venendo in su. Et poi tu, per regola generale, caverai sempre di tucta la somma 3500. Et el rimanente ti mostrerà l'ordine che le migliara sarà el numero delli homini, le centinaia sarà el numero a qual mano lui ha l'anello, le decine saranno el numero delli dita et a qual dito lui lho ha. Et l'ultimo numero sarà el numero delli nodi, a quanti nodi lui lho harà. Siché nota lo exemplo: pognamo l'habbi el quarto homo nella dextra et al tertio dito et al tertio nodo. Hora dirai che doppi el numero delli homini, et farà 8 et agiungha 5 farà 13, et multplichi per 5, farà 65. Hora, a questo agiungha 10, farà 75, in su questo pongha el numero della mano, et havendo nella mano dextra vi porrà su 2 et fa 77 et multiplichi per 10, fa 770, che a questo agiungha el numero delli dita
99r che lui lho ha, che havemo dicto l'ha nel tertio dito, | che farà 773. Et questo multiplichi per 10, fa 7730. Et a questo dirai che ponga el numero delli nodi che sono 3, fa 7733. Hora di questo dico che ne cavi, ovvero facci cavare a lui 3500, cavato di 7733 3500, resterà 4233. Et le 4 migliaia ti denotano el numero della persona che lo ha, et le centinaia ti denoteranno la mano. Et le 3 dicine ti denoteranno le 3 dita et la ultima figura, cioè l'ultimo 3, ti denota el numero delli 3 nodi che lui ha. Eccho che tu dirai che l'anello l'habbi el quarto homo nella mano dextra al dito del mezo all'ultimo nodo, cioè al tertio nodo. Et sebene fussino 100 o 1000 homini, sempre la troverrai. Et così, quando ti fusse facto una nacta, cioè che tu havessi smarito qualche cosa, o uno anello, o uno fazoletto, o simili cose, potrai dire alla brighata: "horsù se io per via de' numero la ritrovo, renderetemela voi" ché questo caso alle volte mi è incontrato a me et mai ti fallirà. Hora nota una altra demostratione bella et gentile a questo medesimo caso acomodata.

Demostratio XXIIIIa a trovare uno anello adosso a uno 99v

Uno altro modo bellissimo da trovare uno anello adosso a uno, et habbilo dove si vole. Prima a te bisogna, a quello che fa el giocho darli ad intendere bene omni cosa, acciò lui non ti faccia errare, ché errando lui erraresti tu anchora. Et perhò, dirai che tucti li luoghi delle persone dove possono nascondere questo anello, o grosso, o ducato, o quello si sia, li vuoi porre uno nome. Et perhò porrai in prima facendoti dal capo. Porrai alla piegha della beretta numero uno et alla beretta dentro, se è doppia, ovvero sotto al capo in sul cocuzolo dua. Et poi fra li capelli 3, et poi in seno 4, et nella scarsella 5. Et poi ad omni dito della mano el suo nome, cioè dirai al dito minimo della mano sinistra 6, et allo anulare 7, et al medio 8, di mano in mano. Et poi, seguendo al pollice di dicta mano 10, et allo indice della mano dextra 11 et continuando sino al minimo della dextra,
chiameralo 15. Poi potrai scendere alli dita delli piè, | ovvero al ginocchio dirai 100r
16, et alle gambe 17, et alle scarpette 18, etc. Potrai fare adosso a uno 20, 30, 40 o 100 luoghi, et a tucti potrai porre suo nome. Et che sia noto a colui che guida el giocho, cioè a colui che multiplicha, et facto et aseptati che saranno a sedere et tu dirai a colui che anoveri da sé sino a quello che ha lo anello. Pognamo sia 6, dirai che lo adoppi, fa 12. Hora dirai vi pongha su 10, fa 22, multplichi per 10, fa 220 et a questo poi dirai che agiungha 5 fa 225, multiplichi per 5 farà 1125. Et a questo dirai che agiungha el numero del luogho dove lui lo ha. Che diciamo sia nel minimo dito della mano dextra che è 15, che giuncto 15 sopra 1125, farà 1140. Hora questo ultimo multiplicharai una altra volta per 10, fa 11400. Et di questa ultima somma, sempre per generale regola, caverai 5250. Et quello ti resterà le migliara ti rapresenteranno el numero della persona che ha la cosa nascosta. El
restante leverai via el zero, cioè taglieralo via. | Et el remanente sarà el luogho, 100v
cioè el numero del luogho, comme in questa. Dinota che cavato 5251 de 11400, resterà 6150 che le 6 migliara ti mostrano che el sexto homo ha la cosa nascosta. Et restati 150, dove taglierai el zero così: 15[o. Restati 15 netto per il luogho dove lui ha nascoso l'anello, o ducato, etc. Et mai ti fallirà questa regola. Et potrai ritrovare una carta, ovvero dire: "io ho tocco uno et voglio che tu mi sappi dire chi io ho toccho et in che luogho" che sia mirabile et di grande ingenio. Et colui che fa il gioco bisogna sia bono abachista, ma nota che multiplicharai senza troppa fatigha, ché omni volta che tu hai a multiplichare per 10 uno numero, basta che solamente vi agiungha uno zero, cioè uno o. Et sarà facto comme se tu havessi a multiplichare 25 per 10, agiugnerai a questo 25 uno zero, farà 250. Et fia facto. Et se tu havessi a multiplichare per 20 25, multiplicharai 25 per 2,
farà 50 et a | questo agiugni uno zero, farà 500. Et così per 30: multiplicharai 101r

---

31 di grande] F di grandissimo

per 3 et poi vi agiugni uno zero. Et fia facto. Et così per 4, 5 et 6, etc. Et se tu havessi a multiplicare uno numero per 100, agiugnerai al numero che tu hai a multiplichare 2 zeri. Et sarà facto. Et così per 1000 agiugni 3 zeri et fia facile et farà tucti questi giuochi havendo questa praticha.

Demostratio xxxa a trovare uno numero che uno havessi in mano senza interrogatione

Che uno piglassi uno numero di grossi quatrini o quartaruoli et simili in mano, etc. Et tu volessi investigare quanto lui habbi preso senza fare altrimenti multiplichare senza domandarlo di niente, darotti uno facilissimo et breve modo quale fia questo: porrai una quantità di quatrini et grossi in tavola et dirai a uno ne pigli una quantità in mano in questo modo che ne pigli con 2 mani et che pigli egualmente tanto da una mano quanto dall'altra, et che operando comme le dirai, li saprai dire quanto lui habbi in mano. Et farai così: dirali che ne prenda
101v quanti lui vole di una mano et mes|coli in l'altra et a te dica quanti ne mette, ovvero tu dirai toglie 8 di questa mano et metteli in questa altra. Ma più bello sarà tenuto se tu li dai libertà di torne quanti lui vuole, ma che ti dica quanti. Et facto che lui harà questo et tu dirai anoveri el rimanente. Et quello che è, dirai li getti via et non ti dica quanto è, ma che lo metta in tavola in modo che tu non habbi notitia di quanti sieno. Et poi li dirai che pigli altretanti di quelli dal'altra mano quanti lui ne gittò in tavola et similiter li metta in tavola. Et poi tu li saprai dire el restante apunto quanto è, che absolute li dirai. Et tu hai in mano tanto. Et nota che sarà sempre el doppio di quanto li facesti mettere di quelli del'altra mano, perché el resto tu li farai gittare via, comme vedi. Eccho lo exemplo più chiaro. Sia che lui habbi preso 16 per mano. Et che pigli 10 di una mano et mettali nell'altra. Et in una mano ne resta 6 et in l'altra ne harà 26. Hora dirali che
102r il resto di quella mano che lui cavò li 10, li getti in tavola et che prima | lui da sé li conti. Et poi dirai che pigli altri tanti di quella altra mano et medesimamente li dirai li getti via. Hora tu sai che in questa altra mano ne haveva 26, tractone 6 resta 20 che apunto è el doppio di quanto vi misse prima della altra mano. Ma per coprire sua facilità potrai fare così: comme lui ha preso, immediate tu potrai mettere in una delle sue mani una quantità a te nota, comme 6, 8 o 10, etc. Et poi nota che tu non facci cominciare a cavare di quella mano dove tu hai messo la tua quantità, perché ti guastarebbe el giocho. Et perhò dirai che della tale mano, nominandola, cavi una quantità a suo modo et mettala nella altra. Et dicati quanto vi mette o cava. Et dirali che cavi di quella mano che tu non hai messo la tua quantità. Et poi dirai similiter che getti via el restante et poi altre

3 Et sarà facto. Et così per 1000 agiugni 3 zeri] F *omissis*

tanti della altra mano. Et tu dirai che a llui resti in quella mano el doppio della quantità che lui rimise della altra mano et più quello che tu vi havessi messo, ché se tu vi harai messo 8 et li resti 20, dirai habbi 28. Po|trai anchora mettere 102v
poi che lui harà gittato via tucto. Et questo si fa perché si copra la sua facilità, perché li astanti, essendo ingeniosi, potrebbono conphrendere sua divisione. Et per questo si dà questa coperta comme vedi.

Demostratio XXXIa a trovare 2 numeri divisi in 2 mani ovvero fra dua persone

Per uno altro legiadro modo potrai pervenire alla notitia di 2 numeri divisi in 2 mani, ovvero fra 2 persone, ovvero in 2 borse o simili. Et è differente dalla superiore dimostratione et non tanto conphrendibile, perché si fa per via di multiplichatione. Ma alli ingnoranti pare più stupenda et più mirabile la superiore che questa, perché in questa ci va qualche interrogatione et nella superiore non si vede che sia demostratione alcuna de interrogatione. Ma questa in sé è più degna regola ché qui va per forza ordinaria et naturale, comme vedrai. Proporrai, dunque, el caso in questa forma: dirai a uno che pigli in dua mani due numeri, uno per mano, et con questa legge che l'uno delli numeri sia pari | 103r
l'altro caffo. Et sieno di che qualità et quanti si vogli et in questi numeri non sia rotto. Ovvero dirai a dua piglino uno numero per uno a te noti, ma non sappi chi li habbi. Comme dire: "dividete 4 et 7 fra voi", o 5 et 10, o 6 et 9, a tuo libito. Et tu voglia ritrovare chi di loro habbi l'uno et chi habbi l'altro. Farai così: dirai a uno di loro che adoppi el suo numero, cioè che lo multiplichi per 2, et all'altro dirai lo multiplichi per 3. Et poi dirai che le loro multiplichationi acozino insiemi et non ti dichino quello fa el tucto, ma bene ti dichino se questa somma è pari o caffo. Et con questo solo saprai giudichare chi ha l'uno et chi ha l'altro. Et nota che se loro ti dicano la lor somma essere caffo, tu dirai che colui che doppiò el suo numero ha el numero pari et quello che multiplichò per 3 ha el numero caffo. Et se loro dicessino loro numero essere pari, allora tu dirai el contrario, cioè colui che doppiò havere el numero caffo et quello che multiplichò per 3 harà el numero pari. Eccho lo exemplo: sia che tu habbi posto in tavola 2 monti di quarteruoli: uno caffo et l'altro | pari, ovvero sarà più bello se con due carte, 103v
ponendole in tavola, ovvero 2 germini o trionfi o simili. Et dica chi vole le pigli. Et bisogna che tu habbi l'occhio ché quando poni le carti o trionfi che l'una sia pari, l'altra caffa, altrimente la regola non ti servirà. Et posto che tu harai le carte et dua le habbino prese che sia che tu ponessi 5 bastoni et 8 coppe a te note, dirai che uno di loro, qual vuoi non inporta, multiplichi el suo numero per 3. Sia quello delle 8 coppe che multiplicato per 3 fa 24. Et al'altro che ha 5

---

17 si vogli] F si vogliono 20 chi habbi] F chi di loro habbi 32 trionfi] F trionfo

bastoni dirai che adoppi el suo numero, farà 10. Hora dirai acozi tucto insiemi, farà 24, et 10 fa 34. Hora dirai ti dichino se la somma è pari o caffo, che in questo caso diranno essere pari. Et tu dirai che colui che doppiò el suo numero harà el numero caffo et colui che multiplichò per 3 harà el numero pari, comme vedi. Et se quello de' 5 bastoni, cioè el primo, havesse multiplicato el suo per 3, fa 15, et l'altro havessi adoppiato el suo, fa 16. Et acozato insiemi 16 et 15, fa 31 et direb- 104r bono essere la somma loro caffo. In questo caso tu giudicherai el cont|rario, cioè dirai che colui che adoppiò el suo numero ha el numero pari et colui che multiplichò per 3 ha el numero dispari. Perhò bisogna che tu noti quando uno doppia o multiplicha per 3 chi lui è, et tenerlo a memoria ché poi non habbia dire chi doppiò ha el numero pari et chi multiplichò per 3 ha el caffo, ma ché tu possa dire expresse: "tu hai el numero pari, cioè le 8 coppe et tu è 5 bastoni". Et nota che questa forza nasce da questo, perché omni volta che tu multiplichi uno numero caffo per uno numero pari, el numero caffo in quello caso diventa pari, ché multiplicha 7 numero caffo per 4 numero pari fa 28, et 28 è numero pari et immobile. Et perhò, multiplicha el numero pari per quanto voi, sempre tornerà pari. Et perhò, in questo caso se tu ti abatti a fare multiplichare a colui che ha el numero dispari per 2, per forza diventerà numero pari, et l'altro che ha el numero pari multiplichi per quanto vuole, sempre starà pari, perché el numero pari è immobile et mai diventa caffo. Nel multiplicare, multiplichalo quanto vuoi, ma el numero caffo fa secondo che lui è multiplichato, che multi- 104v plichi el numero caffo per uno numero pari | torna pari, comme è dicto. Ma se tu multiplichi el numero caffo per numero caffo, torna caffo medesimamente. Et perhò, facendo queste multiplichationi et facendo multiplichare el numero caffo per pari, tornerà pari. Et acozata con l'altre multiplichationi che di necessità fia pari, et mostrerà tucta la somma pari et tu dirai che chi haveva el numero disparo multiplichò per 2. Et se ti mostra caffo et tu saprai ch'el numero caffo sarà stato multiplichato per 3. Et con questa regola tu vieni a questa notitia. Et comme tu hai dal 2 et dal 3, così potrai dare di 4, et di 5, et di 7, et 6, et di 9, et 10, per mutare regola. Et così per 15 et 20 comme lui pigla 2 numeri, uno paro l'altro caffo, così tu piglia dua multiplichationi, l'uno paro et l'altro caffo. Et sieno di che qualità o quantità si voglino pure che non sieno rotti, ché se l'uno fussi 3 et l'altro 1000, ti riuscirà sempre la regola. Et con questo ordine ritroverrai 2 cose: dua anelli differenziati, uno diamante et uno rubino, ponendo ad omni uno di 105r quelli uno nome al'uno pari et l'altro caffo, et a ducati di dua conii et | simili. Et potrai tu porre dua monti di quatrini in tavola a te noti, et l'uno pari et l'altro

1 acozi] F acozino 6 16 et 15 fa 31] F 15 et 16 fa 31. 15–16 et 28 è numero pari et immobile] K et 28 numero pari et el numero pari è immobile 22 pari] K, F inpari 29 dare] F dire

caffo. Et dirà a dua che li piglino. Et guarda di non fare a vedute le brighate che tu facci differenctia al pari o al caffo, ché lo potrebbono conphrendere. Et potrai fare dextramente in varii modi comme tu hai inteso; potrai dare la libertà del loro multiplichatore, dicendo: "multiplicha tu per quanto vuoi", ché al primo non t'inporta. Et poi al secondo dirai tu il multiplichatore che sia in qualità contrario al primo, cioè s'el primo è pari et tu dirai al'altro caffo. Et comme so che col tuo gentile spirito saprai fare, et così potrai dire che battezzino le cose una volta a loro modo et una batezerò a mio. Et potrai per diversi modi ritrovarle, ché dando la libertà comme sopra habbiamo dato, sarà tenuto mirabile dalli ingnoranti che non hanno così vera notitia della forza del numero quanta sia, ché a considerare sua forza et sua virtù sono infinite et miraculose. Et non è scienctia che facci più utile et honore a uno che fa questa dignissima scienctia | mathematica. Et di questo habbiamo dicto a sufficientia. 105v

Demostratio XXXIIa a trovare 3 cose fra 3 persone diverse

E anchora si fa uno altro giuocho bello quale è questo: uno pone in tavola tre monete diverse, cioè uno ducato, uno grossone et uno quattrino, et tre persone fra loro le stribuischono. Et tu vorresti investigare chi di loro le havessi punctalmente, cioè chi el ducato, el grosso, el quatrino. A questo giuocho vi sono molti modi de' quali ne indurremo qui alcuni. El primo fia questo: porrai le monete in tavola et dirai a 3 persone che le piglino. Et comme le hanno prese et tu anovera 24 quattrini, o carte, o quarteruoli. Et di questa ne dà al primo uno, al secondo 2, al tertio 3. Et in tucto ne hai dati 6, et 18 te ne resta in mano. Reponi quelle 18 in tavola et comanda che chi ha del'oro, cioè la prima moneta, ne piglia altre tante quante lui ne ha in mano, cioè se ha in mano una, ne pigli una altra, se ha 2 ne piglia altre 2. Et poi dirai chi ha lo ariento seconda mo|neta, cioè il grosso, 106r
ne pigli dua tanti che lui ha in mano, ché se ne ha 2, ne pigli 4 del monte che lassasti in tavola, et se ne ha 3, ne pigli 6. Et poi dirai che chi ha el quattrino tertia moneta, ne pigli 4 volte tanti che lui ne ha in mano: cioè se lui ne ha una, ne pigli 4 del monte, et se ne ha 2 ne pigli 8, et se ne ha tre ne pigli 12. Et così facto tu noterai quanti quattrini o carte che rimanghino in tavola delle 18 che vi lassasti, et secondo quelle che rimanghano giudicherai. Et possano rimanere in 6 modi, cioè pò rimanere uno et 2, et 3, et 5, et 6, et 7. Et in questi 6 modi, per saperle ritrovare, ti darò 6 parole le quali ti denoteranno per li 6 modi che possano venire. Omni una harà el suo numero di prese, comme vedi qui da parte signato in figura le quali parole sono queste: OAC, AOC, OCA, ACO, COA, CAO. Con queste 6 parole retroverai li 6 numeri che possono venire ché se avanza

---

12 et honore] F et più honore 15 quale è questo] F *omissis* 18 chi el ducato] F chi ha el ducato 28 cioè se lui ne ha una] F se lui ne ha in mano una 35 oca] K oha

una et tu piglierai per tua guida OAC. Et se avanza 2 et tu piglierai AOC. Et se
106v avanza 3 et tu pi|glierai OCA, avanzando 5 piglia ACO. Et se avanza 6 piglia COA.
Et se avanza 7 pigliarai CAO. 4 non ti può mai avanzare, comme experimentando proverai. Et perché tu intenda el significato di queste parole: per lo O tu intenderai l'oro, per la A tu intenderai l'ariento, per lo C intenderai el rame, cioè per lo O el ducato, per la A el grosso, et per lo C el quattrino. Dunque, quando avanzerà una tu piglierai OAC et dirai che colui che desti uno harà el ducato, et chi desti 2 harà el grosso, et chi desti 3 harà el quatrino. Et se avanza 2 el primo harà el grosso, el secondo el ducato et il tertio el quattrino. Et se avanza 3 dirai OCA, ché el primo ha el ducato, el secondo el quattrino, el tertio l'ariento. Et se avanzerà 5 dirai ACO, ché il primo harà el grosso, el secondo el quattrino, el tertio el ducato. Et se avanzerà 6 dirai COA, ché il primo ha el quattrino, el secondo el ducato, el tertio el grosso. Avanzando 7 dirò CAO, ché il primo ha el quattrino, el secondo el grosso, el tertio el ducato. Exemplo: diciamo sieno 3 che habbino
107r nome Antonio,| Bartolo et Carlo, et che Antonio habbi el grosso, Bartolo el quattrino et Carlo el ducato. Et diciamo che a Antonio habbi dato uno, a Bartolo 2, et Carlo 3. Et dirai che chi ha el ducato ne pigli altretanti che lui ha in mano, che lo ha Carlo che ne ha 3 in mano, dunque ne piglerà altre 3 del monte delle 18, ne resterà 15. Hora dirai che chi ha l'argento ne pigli dua tanti che lo ha in mano. Antonio, che ha uno in mano, che ne piglierà 2 del monte et resteranno 13. Et poi dirai a chi ha el quattrino, ne pigli 4 volte tante. Et hallo Bartolo, che ne ha in mano 2, dunque ne piglerà 8 del monte, et resterà 5. Et omni uno ha preso. Dunque tu vedi che nel monte resta 5, che in questo caso hai a pigliare, comme vedi in figura, ACO, che ti mostra che Antonio prima ha el grosso, Bartolo secondo ha el quattrino, Carlo tertio ha el ducato. Et così ti riuscirà di tucti e' modi che ti potessi venire, ovvero che loro potessino pigliare comme potrai per te medesimo experimentando provare. Et nota che queste seguenti parole
107v scripte qui in margine faranno el medesimo | effecto che le sopradicte che sono queste: "Mathei Beati Marie Deitas dignare pietas" ché Mathei è in luogho di OAC et Beati è in luogho di OCA, et a luogho per luogho. Ma nota queste lettere vocali si pigliano, che la A si piglia per el ducato, cioè per la prima cosa. Et lo E si piglia per lo ariento, per la seconda cosa. Et lo I si piglia per lo rame, per la tertia cosa. Et così potrai fare li medesimi effecti con uno altro ordine che si dà, che sono dua versi che dicano così:

Asequor unum duos belandos tresque latrones.
Dogmate sex ocreas dat septem quinque reportant.

10 dirai oca] K dirai ocha

In questi dua versi ti denota quanto omni figura rapresenti. Et per tanto la lettera A dinota lo oro, et la lettera E denota l'ariento, et la lettera O denota lo rame. Et così potrai per queste tre vie che ti ho mostro, pervenire alla prefata cognitione brievemente che fia bella et habbi sempre l'occhio a chi tu dai la unità et a chi tu dai el 2 et a chi el 3. Et mai ti fallirà. Hora nota una altra via pure alla medesima cognitione acomodata et piaceratti senza dubio.

Demostratio XXXIIIa al medesimo giuocho regula 108r

Per una altra via potrai alla predicta notitia pervenire. Et fassi così: sieno pure 3 che habbino el ducato, el quattrino et il grosso, et tu darai a uno di loro 12. Et questo a chi tu dai 12 si chiami primo, al'altro darai 24 et chiamasi secondo, al'altro darai 36 et chiamasi tertio. Et poi dirai a colui che ha el ducato gitti via la metà di quanto lui ha, et a colui che ha el grosso gitti via i 2/3 di quanto lui ha. Et colui che ha el quattrino, getti via le 3/4 di quanto ha. Et poi dirai che tucte le lor somme che li restano acozino insiemi et che ti dichino quanto resta in tucto. Et nota comme nella superiore che questa somma può restare in 6 modi, cioè in 23, in 24, in 25, in 27, in 28, in 29, che in 26 non può mai restare. Comme potrai tu vedere, in questo caso, comme nella superiore, ti bisogna adoperare le sopradicte parolle et versi, ché, quando ne avanza 23, tu piglierai la prima posta, cioè OAC, ovvero Mathri, non fa caso ovvero asequor. Omni uno ti mostrerà uno medesimo effecto. | Comme se tu havessi dato ad Antonio 12, a Bartolo 108v 24, a Carlo 36. Et che il primo habbi el grosso, el secondo habbi el ducato, el tertio el quattrino. Dirai che chi ha el ducato gitti via la metà di quello che ha, che in questo lo ha Bartolo che ha 24, gittando via la metà li resta 12. Et poi dirai a chi ha l'ariento gitti via la 2/3 de la quantità che si trova, che lo à Antonio che ha 12 in mano, ché gittando li 2/3, gitterà via 8 et resteralli 4. Et poi dirai che chi ha el quattrino gitta via li 3/4 di quello che si trova, che el quattrino l'ha Carlo che li desti 36 in mano. Dunque, gittando via e' 3 quarti, getterà via 27 et resteralli 9 in mano. Hora dirali che acozi tucte queste somme li restano, cioè 12 che resta a Bartolo, 4 che resta ad Antonio, et 9 che resta a Carlo, faranno in tucto 25, che hai a pigliare la seconda figura, comme vedi qui signato da parte, ché questa ragione non va comme la superiore comme vedi ordinata, et perhò si muta la quantità, comme vedi che è qua in margine signata. Ché a 23 piglierai | 109r OAC, a 24 piglerai OCA, a 25 AOC, a 26 non rimane mai, ma a 27 piglierai COA, a 28 ACO, et a 29 CAO. Et sempre ti riuscirà comme tu medesimo potrai provare. Et per consequenter si mutano le altre, cioè Mathri Beati Marie muterassi in questo secondo modo et dirai così: Matri Marie Beati dignare Deitas pietas. Et

8 predicta notitia] F medesima notitia

conseguentemente si muterà Axsequor et li altri nomi, comme vedi. Et mai ti fallirà questa regola, comme tu medesimo potrai provare, ché in tucti e' modi ti riuscirà.

23	25
OAC	AOC
24	28
OCA	ACO
27	29
COA	CAO

Demostratio XXXIIIIa a fare el medesimo effecto per via di numeri et multiplicationi

Per via di numeri et multiplicationi, potrai venire al medesimo effecto senza queste parole o versi. Et fassi in più modi per una di multiplicationi. Prima lo potrai fare in questo modo: darai ad omni uno uno numero a tuo modo, cioè al'uno 2, al'altro 3, al'altro 5. Et poi dirai a uno di loro multiplichi el suo numero per 2, et sia el primo. Et al secondo dirai che multiplichi el suo numero per 109v 11. A l'altro, cio|è al tertio, dirai che multiplichi el suo per 12. Et tucte queste 3 somme l'acozino insiemi. Et poi tu da te dirai: "io ho dato al'uno 2, et al'altro 3, et al'altro 5, che in tucto fa 10 puncti". Et questo multiplicherai via 12, cioè via 6, numero che multiplichò l'ultimo. Et di questa tua multiplicatione dirai ne cavi la lor somma, et quello resta partirai per 2 mancho, che l'ultimo quale fu 12, cioè per 10, et quello ne viene sarà el numero che facesti doppiare. Et quello che avanza sarà el numero di colui che multiplichò per 11, et l'altra di necessità sarà el restante. Hora sia comme è dicto: che Antonio habbi 2, et Bartolo 3, et Carlo 5, che fra tucti hanno 10. Poi dirai Antonio che adoppi el suo, fa 4, et poi a Bartolo che multiplichi el suo numero per 11, fa 3 via 11, 33. Et al'altro dirai multiplichi per 12, fa 5 via 12, 60. Hora dirai acozino 4, 33 et 60 insiemi, farà tucto 97. Et tu multiplicharai l'ultimo multiplichatione, cioè 12, per tucta la somma che hanno fra loro, che è 10. Hora multiplicha 10 via 12, fa 120. Et tu 110r dimanderai che differentia sia dalla tua somma alla loro, che | si vede cavando la lor somma dalla tua, ché traendo 97 di 120, resta 23, et tanto è la differentia. Et per non li fare acorti del facto, basta che dica ti dichino la differentia che è dalla tua somma alla sua, che comme habbiamo decto è 23. Et questo dica che

11–12 tucte queste 3 somme] K tucta questa tre somma 13 al'altro 5] F al'altro 4 parta per 2 meno ch'è l'ultimo tuo multiplichatore ché l'ultimo multiplichatore è 12. Dunque, partendolo per 2 mancho, lo harai a partire per 10, ché partito 23 per 10 ne viene 2 per el primo numero et avanza 3 per il secondo. L'altro di necessità fia el resto fine in 10, che fia 5. Et così potrai ritrovare per questa regola che la prefata regola dissi di sopra nella quarta dimostratione, dove è regola generale a trovare 3 parte di uno numero diviso in 3 modi. Et per tucti 3 li modi potrai ritrovare questo medesimo giucho che se bene me ricorda questo debba essere uno de quelli, perhò non mi distenderò in questo capitolo. Andrai a leggiere quella dimostratione et troverai più difusamente le regole generali date, a trovare 3 parte di uno numero diviso, et 4 parte et 5 parte di uno numero diviso et 6, 7, | 8 et mille, se 1000 fussino le parte. Siché potrai ritrovare 3 cose, 4 cose, 110v 5, 6, 10, 20, secondo che bisogna, ché nella sexta dimostratione di questa parte seconda vi si dà regola generale a trovare quante parti tu volessi di uno numero diviso, pure che del numero diviso non sia nota sua quantità et in quante parte è diviso. Perhò non mi extenderò più in questa dimostratione.

Demostratio xxxiiiiia (sic!) al medesimo effecto fare per altro modo variato
Per via di numero potrai anchora per uno altro breve modo pervenire alla prefata notitia. Comme a dire: dirai ad omni uno per più libertà che pigli uno numero a suo modo. Et che questo tale numero non passi 9 et che non pigli uno medesimo numero, uno che l'altro, ma diversi et vari. Et sia che l'uno habbi preso 3, l'altro 5, l'altro 9 et che poi habbino preso omni uno la sua moneta a te incognita, cioè che tu non sappi specificare chi habbi l'una et chi habbi l'altra. Et per regola generale dirai a chi ha el ducato radoppi la sua somma | et poi vi 111r pongha su 5 in sul doppio che lui ha et poi el tucto multplichi per 5 et di poi vi pongha su 10. Et poi vi ponga su el numero di quello ha el grosso et questo multiplichi per 10. Et a questa somma ponga el numero di colui ha el quatrino. Et poi dirai che di tucta la somma ne cavi 350 et quello resterà. Li centinara saranno el numero di quello che ha el ducato et le decine saranno el numero di colui ha el grosso et l'ultima figura sarà el numero di colui che ha el quattrino. Sia che Antonio habbi 3 et Bartolo 5 et Carlo 9. Et pognamo Bartolo habbi el ducato, Antonio el grosso et Carlo el quatrino. Dirai chi ha il ducato radoppi el suo numero. Hallo Bartolo che ha 5, fa 10, et a questo dirai agiungha 5, fa 15. Hora questo si multiplicha per 5, fa 75 et a questo agiugni 10, fa 85. Et a questo agiugni el numero di colui ha el grosso et chi ha el grosso ha 3, fa 88. Et questo multplicha per 10, fa 880. Et a questo agiugnerai el numero di Carlo che ha il quattrino che ha 9, farà 889. Et di questo, dico, cavi 350, che tracto 350 di 889 | 111v resterà 539. Et questo ti denota che a chi tu desti di principio 5 harà el ducato, ché desti 5 di principio a Bartolo. Et a chi desti 3 harà el grosso, et a chi desti 9, harà el quattrino. Dunque Bartolo ha el ducato, Antonio el grosso, Carlo el qua-

trino. Et sempre ti riuscirà. Et questo è per la regola delli 3 dadi di sopra data, pure ch'el numero che tu dai loro non passi 9 et habbino puncti variati, cioè che non habbino uno medesimo puncto. Et così per quella regola dello anello potrai ritrovare 4 cose fra 4 persone divise, purché li numeri non passino el 9, comme nella dimostratione 29 di questo libro puoi vedere. Et mille belle gentilezze si potranno fundare in su questa data regola, che a me saria fatidioso tucto potere recitare, ma questa basta alli ingenii acutissimi.

Demostratio XXXVIa a trovare una carta fra 16 pensata

Uno altro giuocho de industria et piacere si poterà fare per via di linea recta.
112r Comme se tu havessi facto di 16 carte, o di 16 grossi, | o ducati, o quatrini,
purché siano differentiate et diverse, perché non potria l'occhio tanto apertamente judicare, comme s'elle fussino tucte d'uno conio, ma di diversi conii et stampe. Ma a questo giocho ti serviranno meglio 16 carte, ovvero 16 schacchi diversi, cioè mescolati bianchi et neri che altro. Et questi 16 scacchi, o carte, o grossi, etc., farali porre in su una tavola distesi a filo, comme vedi qui da canto, per figura. Perché acioché meglio possa conphrendere, porremo 16 figure della A, B, C, comme vedi. Et poi tu, per volere ritrovare dicta littera che lui ha toccha, farai così: dimandali in quale è delli dua fili. Il che saputo et tu comincia sempre, per regola generale, da quello filo che non è la sua che lui ha toccho, et piglia una di quelle et una dell'altro filo di quelle che toccho. Et piglia in questo modo: cioè piglia una carta del filo che non ha toccha et porrala in tavola coperta in modo che la dipinctura di epsa carta vengha in verso la tavola et di
112v sopra sia la sua coperta. Et poi piglia una altra charta del filo che lui prese | et
mectala sopra quella nel medesimo modo et poi l'altra, medesimamente, ritornando al filo dove non toccho, in modo che te venghino una d'uno filo et l'altra dell'altro alternatim. Et comme tu le hai levate tucte 16 et aseptatele nel modo sopradicto, et tu ordinerai 2 altri fila, mectendole ordinate comme le sono. Et per questo fare, volterai le carte sotto sopra, cioè cominciando ad aseptare el filo da quella che prima levasti. Et quando tu asepti li fili, non le mectere alternatim l'una nell'una et l'altra nell'altra, perché non faresti el tuo intento, ché per variare le monete et carte tu vieni alla conclusione di epsa, et perhò comincia uno filo et non restare se tu non hai aseptate tucte 8. Et comme hai finite et tu comincia l'altro filo et finisci tucte le altre 8 et harai diposti due fili che vi sarà per uno 4 dal'uno et 4 dal'altro. Et nota che donde tu cominci racorre, non hai a cominciare a riporre. Comme se tu cominci racorre da piè, non ti fare a
113r riporre poi da piè, ché non riuscirebbe el facto. Et | perhò, comme una volta le
hai ricolte et riposte nel modo sopradicto, dimanda una altra volta in qual filo è

14 neri] F nigri 18 fili] F file

la cosa pensata, il che saputo una altra volta, racorrai le carte nel modo predicto, cominciandoti sempre nel filo che non è la carta presa, et ponendola in tavola coperta et poi pigliandone una altra dell'altro filo et ponendola nel medesimo modo sopra quella. Et per fine che tucte le harai una altra volta racolte, et poi ordina 2 altri fili nel modo predicto, cominciando al contrario a porre che non cominciasti a ricorre, et ponendo prima uno filo intero et poi l'altro finché li harai finiti tucti dua. Poi dimanda una altra volta in che filo è la sua carta presa. Et saputolo una altra volta, racorrai tucte nel modo dicto, cominciandoti sempre dove non è la pensata et racogliendo alternatim una del'uno filo et l'altra del'altro. Et comme le harai ricolte tucte et tu ordinerai nel medesimo modo e' tua fili. Et comme li hai aconci, dimandirai la pensata in qual filo è et in quello filo che è sempre sarà la tertia a contare onde ti cominciasti a ra|corre et non 113v
dove cominciasti a porre. Hora ecchoti lo exemplo: habbi 2 fili o linee che sia la prima A, B, C, D, E, F, G, H et il secondo fia K, L, M, N, O, P, Q, R, comme qui da canto li vedi dispositi.

A	K
B	L
C	M
D	N
E	O
F	P
G	Q
H	R

Et sia che colui habbi preso L et tu comincia a ricorre et comincia dal filo dove non è che harai a cominciare a raccorre da piè, perché hai cominciato a porre da capo et dirai: "per il primo filo hai a fare farai comme vedi qui in figura: H, R, G, Q, F, P, E, O et per il secondo filo farai D, N, C, M, B, L, A, K".

H	D
R	N
G	C
Q	M
F	B

7 filo] K filio

P L
E A
O K

Et hora te farai dire in quale filo è la sua carta pensata che dirà nel filo del D, N, C, etc. Hora tu raccorrai et ordinerai 2 altri fili nel modo predicto che saranno li dua altri fili: O, R, E, A, P, L, F, B, et l'altro sarà: Q, M, G, C, R, N, H, D.

O Q
R M
E G
A C
P R
L N
F H
B D

Et poi dimanda una altra volta in quale filo è la sua carta pensata che ti dirà è in quello della O, R, etc. Et ordinerai 2 altri fili racogliendoli nel modo predicto, cominciando da quello dove non è, comme te ho mostro, et harai 2 altri fili in questo modo, cioè: D, B, H, F, N, L, R, P, et l'altro sarà C, A, G, E, M, R, Q, O.

D C
B A
H G
F E
N M
L R
R Q
P O

114r Et | a questa volta domandandoli donde o in quale linea sia, che dirà: in quella di D, B, etc. Et tu contando da piè, perché hai cominciato racorre da piè, dirai che

8–9 in quella di D, B, etc.] F in quella di C, A, etc.

sia la tertia littera di quel filo, che comme vedi è L. Et da te potrai tu in infinito procedere. Et comme qui è decto di 16, potrai di 20 et di 30 etc., che sia bella fantasia. Et perhò non mi distenderò più in questa dimostratione, andaremo al'altra.

Demostratio XXXVIIa delle figure delli pianeti

Danno li antiqui et famosi astronomi alli V pianeti, a ciaschuno una figura quadra, partita in più parte quadre et in omni quadro hanno messo el suo numero. Et perhò li voglio in questa mia operetta recitare, perché sono acomodati a fare molti piacevolissimi giuochi, et per questo si denota che in omni scientia la mathematicha è necessaria, la quale li astrologi sommamente adoperano in loro propositioni.

Et perhò ad omni pianeto hanno dato la sua mathematicha figura, comme a Saturno hanno dato una figura di 9 quadri, con sua numeri drieto,

Saturno 1

2	9	4
7	5	3
6	1	8
15		

comme vedi qui in figura da parte. Et in omni quadro sono proportionati in 114v
modo che per qualunque verso tu conterai sarà 15. Et el primo filo è, 2, 9, 4, il secondo è 7, 5, 3 il terzio filo è 6, 1, 8. Et contali per che verso tu vuoi, sempre fa 15.

Iove 2

16	3	2	13
5	10	11	8
9	6	7	12
4	15	14	1
34			

6 alli V pianeti, a ciaschuno] K alli V pianeto a ciaschuno, F alli pianeti a ciascuno 9 piacevolissimi] F piacevoli 13 comme a Saturno hanno dato una figura] F *omissis*

Et a Iove hanno dicata la figura di 16 quadri, 4 per faccia et 4 faccie et in omni quadro suo numero. Che il primo filo è: 16, 3, 2, 13 et el secondo è: 5, 10, 11, 8, et come vedi il tertio è: 9, 6, 7, 12 et il quarto è: 4, 15, 14, 1, comme vedi qui in figura. Et anovera per che verso vuoi, sempre farà 34 et per diametro a tuo modo.

Marte 3

14	10	1	22	18
20	11	7	3	24
21	17	13	9	5
2	23	19	15	6
8	4	25	16	12
65				

Et a Marte hanno dato la figura di 25 quadri, 5 per faccia et in omni quadro hanno messo sua numeri, comme vedi qui da piè le figure che per omni verso fa 65. Et sono sua numeri la prima faccia: 14, 10, 1, 22, 18, et la seconda: 20, 11, 7, 3, 24 et la tertia: 21, 17, 13, 9, 5 et la quarta: 2, 23, 19, 15, 6 et la quinta: 8, 4, 25, 16, 12.

Sole 4

1	32	34	3	35	6
30	8	27	28	11	7
20	24	15	16	13	23
19	17	21	22	18	14
10	26[a]	12	9	29	25[b]
31	4	2	33	5	36
111					

a 26] K, F 16
b 25] K, F 15

115r Et al sole hanno dato la figura di 36 quadri, che | omni quadro ha al suo numero et fa, comme vedi qui in figura, per omni verso 111, comme vedi.

Et a Venere hanno dato la figura di 7 case per omni verso, che sono 49 case et in omni casa suo numero, che in tucto fanno per omni verso 125.

Venus 5

4	35	10	41	16	47	22
29	11	42	17	48	23	5
12	36	18	49	24	6	30
37	19	43	25	7	31	13
20	44	26	1	32	14	38
45	27	2	33	8	39	21
28	3	34	9	40	15	46

175

Mercurii 6

4	7	59	60	61	62	2	5
49	15	54	12	53	51	10	16
45	42	22	21	20	19	47	44
32	34	35	29	28	38	39	25
40	26	27	37	36	30	31	33
17	18	46	45	44	43	23	24
9	55	14	52	13	11	50	56[a]
64	63	3	4	5	6	58	57

260

a 56] K, F 66

Comme vedi qui a piè, sono le soprascripte figure disposte a molti giochi dilectevoli. Comme dire: "metti queste poste in tavola et io mi voglio aporre toccandone tu una in mente tua quale tu habbi toccha". Et fassi in dua modi. Et
il primo fia questo, dicendo: | "dimmi in quale fila ella sia per lungho et per 115v
largho" che verbi gratia diciamo lui havere toccho la figura che ha 41, dove di sopra è lo "o". Et tu dimanderai per la lunghezza in qual filo è, che ti dirà nel filo del 21, dove è signato la linea, comme vedi. Et poi dimandando per la largheza in qual filo sieno, ti dirà in quello del 23, dove è la croce così "+". Et
tu dirai che per forza sia dove li 2 fili si contrano insieme che fia nel punto | 116r

4 fila ella sia] F fila la sia

di 41 con l'"o" di sopra, comme vedi signato, ma questo modo fia troppo facile. Ma se tu farai così in uno altro modo, comme lui ha toccho con l'animo el suo numero o posta, et tu ti farai dire in quale filo è la sua posta. Et poi, comme hai saputo, et tu rimuta tucte le poste, cioè ricorrai tucte per lo lungho et poi le riponi tucte per el contrario che tu le hai levate, cioè se tu le hai levate per lo lungho et tu le ristendi per el traverso. Et poi dimanda per lo lungho in quale filo è, et tu vedrai che in quello filo che lui ti dirà essere la sua pensata non ve n'è, se non uno del filo del quale lui prese, et quella dirai sia epsa.

Luna 7

5	54	13	62	21	70	29	78[a]	37
46	14	63	22	71[b]	30	79	38	6
15	55	23	72[c]	31	80	39	7	47
56	24	64	32	81	40	8	48[d]	16
				o				
25[e]	65	33	73	41	9	49	17	57[f]
66	34	74	42	1	50[g]	18	58	26
35	75	43	2	51	10	59	27	67
76	44	3	52	11	60	19	68	36
45	4	53	12	61	20	69	28	77

a 78] K, F 18
b 71] K, F 21
c 72] K, F 22
d 48] K, F 4
e 25] K, F 23
f 57] K, F 51
g 50] K, F 51

Et eccho qui la figura et lo exemplo più manifesto nella figura di Jove che sono 4 case per omni verso. Et dirai che habbi toccho una di decte case. Et sia che lui habbi preso 10 che è nel filo della "+", comme vedi. Hora tu muterai tucte dicte poste aseptandole in figura, comme vedi nella seconda figura che sono
116[v] tucte mutate et stanno al contrario che prima. Hora | tu li domanda in quale filo sia la sua posta pensata, che ti dirà nel filo della + medesimamente. Et tu dirai che sia la seconda di quel filo, contando di sotto, perché tu cominciasti a contare ovvero a levare di sotto, et perché era nel secondo filo. Perhò dirai che sia la seconda casa, cioè la seconda figura di quello filo, a cominciare a contare di sotto. Et così potrai operare con la figura della luna et con tucte.

Iove prima

	+		
16	3	2	13
5	10	11	8
9	6	7	12
4	15	14	1

34

seconda

		+	
1	12	8	13
14	7	11	2
15	6	10	3
4	9	5	16

34

Demostratio XXXVIIIIa a fare con li supradecti instrumenti con le carte

Fassi con carte et con li germini simili giuochi et paiono nel conspecto delle donne cose miraculose. Et quando ti trovi a una tavola che vi sia più brigate et tu puoi torre le carte overo e' germini ché quante più sono le carte et quanto fia maggiore el caso, tanto più sarà stupendo apresso a ciaschuno che di questo giuocho non sarà amaestrato. Et tu potrai pigliare le carte che con epse farai 7 fila, ché con le carte non potrai più farne. Et saranno 7 per filo | ché così 117r
bisogna che sieno, perché tramutandole et non tornando al modo medesimo sarebbe facile a conphrendere sua regola. Et perhò bisogna che sieno tante per uno verso quante per l'altro. Potrai torre li germini, ché con epsi farai 9 fili. Et havendo i germini sarà più bello. Et facto che tu harai le poste et tu facendoti dall'uno delli capi della brighata, dirai a uno, quale in tua fantasia asepterai, che sia primo, che nel primo filo ne pigli una di epse carte, a suo modo, et tengha in sé. Et poi a uno che sia a llato a llui dirai che ne pigli una altra nel secondo filo, et sia questo in mente tua secondo. Et poi a uno che sia a lato al secondo dirai che ne pigli una nel terzo filo, et questo a te sarà tertio. Et poi al quarto nel quarto filo, et al quinto nel quinto, et sino che in omni filo ciascheduno ne habbi preso una. Et nota che per quello verso che tu hai dicto che le piglino, per quello

hai racorre tucte dicte charte. Ché se tu hai dicto che le piglino per lo lungho
117v et tu racogli per lo lungho, filo per filo, facendoti sempre a racorre | nel primo filo. Et se tu hai dicto che piglino per traverso et tu racorrai per traverso. Et poi, comme tucte le harrai raccolte per lo lungho et così habbi facto pigliare. Et diciamo habbi tu 7 fili et non più, et tu di novo piglierai dicte charte, et comincia a riordinare li fila non comme prima stavono, ma in contrario, cioè a stenderli per lo traverso, havendoli racolti comme è dicto per lo lungho. Ma quando li havessi raccolti per lo traverso, ti direi che hora li stendessi per lo lungho. Siché guarda comme fai, perché el giocho non riuscirà altrimenti facendo. Hora diciamo le stenda per il traverso. Et asepterai el primo filo che era per lo lungho, in modo che tucto vengha per traverso. Et così el secondo, el terzio, el quarto, di mano in mano tucti. Et poi comme tucti li hai aseptati, et tu facendoti dal primo, li dimanda la sua carta pensata in qual filo è per lo lungho. El che saputo, dirai che la sua carta sia la prima di quello filo, che non può essere altrimenti, perché tucto el suo filo è il primo di sopra et di necessità in quello che lui ti dice
118r sarà la prima. Hora | vieni al secondo et dimandalo in quale filo sia la sua per lo lungho. Il che saputo, in quello filo che lui ti dirà sia la sua carta. Di necessità sarà la seconda di epso filo, perché el secondo filo traverso è tucto el suo filo. Et poi, al tertio dimandato in quale filo è la sua et saputolo, dirai in quello filo che lui ti mostra ché la tertia carta sia la sua per le ragioni preallegate. Et poi dimanda el quarto in quale filo è. Lui te lo dirà et tu dirai che sia la quarta. Et così al quinto dirai che sia la quinta del filo che ti mostra, et al sexto la sexta, et al septimo la septima. Et nota che tu sia advertito bene in questo domandare la seconda volta, ché tu hai a racorre le carte per quello verso che tu facesti prima pigliare, ché se tu dimandassi per altro verso non ti verrebbe facto in questo el giocho, perché tu sai per il traverso in che filo è quella di ciaschuno. Et perhò bisogna che tu domandi per el contrario di quello che tu sai. Et se tu havessi 2 o 3 paia di germini che tu facessi 100 per omni verso, tanto più bello sarà
118v tenuto, purché la tavola sia capace della quantità delle poste | che tu vuoi mettere. Et mai ti fallirà, tenendo bene questa regola alla memoria, la quale è facile a saperla, ma solum quando fussino assai poste è difficile el tenerle a mente primo, secondo, tertio etc., ma aseptandole ordinatamente fia facile.

Seguita la tertia parte del libro de' giochi mathematici

Ritrovandomi Magnifico Juliano nella stagione che illuminante Apollo con li
sua fulgentissimi raggi comincia a riscaldare li acutissimi corni dello indomito
tauro, et uno giorno fra li altri havendo già la obscurissima nocte dalli miseri
amanti desiderata occupato el nuovo emisperio, io che alla mia rude et inhor-
nata opera ero intento, et per volere a quella dare sua vera perfectione, havendo
già presa la penna et posto la mano in su la candidissima carta et con l'altra
appoggiato el volto substentando el capo per le intrinseche cogitationi, et molti
et vari innumerabili pensieri, defatichati furono in uno instante li mia tenuis-
simi membri da uno leve et suavissimo sonno o|ppressi. Et così stando, mi 119r
pareva essere nel profundissimo et spatioso mare, in su una debilissma navi-
cella collocato. Et stando in epsa, mirando el mare, hinc et inde con suavissimo
et tranquillo vento, ne pigliavo grandissimo piacere a vedere li pesci per lo fluc-
tuante et amplo mare andare li loro amori exercitando, ma la fortuna invida di
tanta mia consolatione, subito mutata la tranquillità in crudelissima tempesta,
prestamente el mio dilecto et consolatione in grandissima mestitia et grave
dolore convertì et el sole lucidissimo divenne obscuro et velato, dalli aquosi
nuvoli coperto. Eolo, inpetuoso Idio delli furibundi venti a Euro et a Noto et
Africo le ferrate chatene relapsante et quelli rapidissimi el placido et tranquillo
mare in maxima fortuna rivoltando, non altrimente che le furie infernali hor
qua, hor là, movendo le timide acque faceva grandissimi gruppi.

Et hora in occidente et hora in oriente quello spingevano, el trombetto Tri-
tone con la sua tremebunda et horribile tromba sopra el suo | carro anunptiava 119v
le furie dello inpetuoso Neptuno, per li quali sopradecti accidenti la mia povera
navicella, in uno tracto perduto el timone et le vele et le sarte, in qua et in
là vagabunda trascorrendo in quelle parte si volgeva dove li inpetuosi venti
la spingevano. Ma per questa tempestosa fortuna non vi era periculo che la
mia barchetta afondassi, perché li primi venti non mi pareva havessino poten-
tia di potere epsa submergere. Pure io, comme quello che di peggio dubitavo,
con le mani giuncte non lassavo decta nessuna che da me non fusse humil-
mente invocata, a quella suplicando che da sì misera fortuna mi liberassino.
Et così stando viddi verso di me venire uno homo nello aspecto molto terri-
bile et porgendomi la mano mi diceva che di niente temessi, ché lui era di tale
auctorità che li venti, el cielo, el mare li rendevono obedientia et promecte-
vami, se a llui volevo prestare fede, di fare cessare subito sì inpetuosa fortuna
et che sicuramente epso nella mia barchetta raceptassi. Et con queste parole,
aproximandosi me|glio, potetti sua immagine vedere. Questo hora era tucto 120r
nero, non altrimenti che li Ethiopi, et haveva una veste lungha fine alli piedi,
di modo che niente di lui si vedeva, et era discinto et el capo suo di una sub-

tilissima tela era tucto avolto. Di che colore la veste et il tegmine del capo fussino, non facilmente cognoscere si poteva, ché a me pareva hor rosso, hor biancho, hor verde, et parevami che in omni momento mutassi colore. Et nella sua dextra mano haveva uno candidissimo hermellino et nella sinistra uno libretto aurato. Io, benché di tale aspecto alquanto temessi, pure le promisse grande che lui mi faceva, me spinsono ad acceptare quello nella mia navicella, spectando da quello salutifera quiete, el quale non prima fu entrato nella mia barchetta che mi dette in mano quello hermellino. Et lui atteso a schacciare le pioggie et li venti, alquanto fece la marittima fortuna quietare, il che fu quasi niente, perché pocho stando el mare che irato era et non contento alle
120v prefate cose, con una furia grandissima incominciato di | novo la sua inpetuosa ira a dimostrare, li dalphini si vedevano assalti solcare le furibunde aque. Onde la mia barchetta, con la quale speravo felice in porto ritrovarmi, dalla fortuna trasportata hor qua, hor là, hor giù, hor su, senza guida alcuna si agirava, et quello che più mi spaventava è che io viddi quattro balene di smisurata grandezza ad epsa acostarsi, et con omni loro industria si forzavano farmi trabocchare. Pensi ciaschuno in che stato el mio tribulato core si ritrovava. Le inpetuose onde da omni parte nella mia barcha abundavano, di modo che grandissimo timore me spingeva a notare l'aque che ad omni hora insurgevano. Et ricordandomi di colui che nella mia barcha pocho innanze avevo riceptato, a llui voltomi, pregavolo che di tal fortuna, comme promesso mi haveva, me liberassi, di che lui altro che confortative parole non mi dava, onde rivolto alli superiori dei, genuflexo a tucti humilmente racomandavami. Et così standomi, in uno instante viddi quello homo gettato giù el suo mantello con il quale sua
121r turpi|dissima et spaventevole forma copriva et omni altra cosa al'uno delli lati della nave apicchatosi, quella con tucta sua forza et ingenio cerchava al tucto afondare et submergere, per il quale accidente fui al tucto privo di potere mai di tanto acerbissimo caso scampare. Et omni altra cura da canto lassata, con uno tronchone di uno remo adesso voltatomi et quello fortemente percotendo me ingegnavo da tale inpresa levarlo. Ma lui di mia percosse niente curava, ma ridendo si mostrava schernirmi et pure ad affondare la barcha era intento. Et haveva el prefato homo el capo et la gola nero, li occhi carboni di ardentissimo fuocho parevano, el busto di serpente et li piedi di naturale cervio si potevano iudicare. Onde, levata omni speranza della mia salute, ad hora ad hora spectavo che la mia nave afundassi et me dalle furibunde onde miseramente essere submerso timevo, niente di mancho sempre allo altissimo celo tenendo li occhi fixi, a tucti li supervi dei devotamente mi racomandavo. Et

5 libretto] F libro 7 entrato] K entrata

mentre che così stavo, subito viddi aparire una candidissima | nube, quasi per- 121v
pendiculare sopra la barchetta con tanto vago splendore che la mia mente al tucto dalla morte che già vicina vedeva opressa fu alquanto da simile luce excitata et levatomi in pié nella mia barchetta, che più che li dua terti era dalle aque occupata, et alzato il mestissimo volto, viddi quella nube verso di me velocemente venire. Donde, presa alquanto di speranza, stavo attonito a quello che di simile spectaculo nascere doveva. Quando factamisi più propinqua, viddi che in decta nube era una spetiosissima donna, la quale era in una sedia di radiante oro hornata et le sue veste erano candidissime non altrimente che le montuose neve et in capo haveva una corona di verde lauro et nella sinistra mano teneva una palla di tucto purissimo oro et nella dextra haveva uno ramo di viridissima oliva et pareva nel suo aspecto di grandissima auctorità. Et alla sua mano dextra in una sedia simile alla sua sedeva uno homo di aspecto grave et somma eloquentia et dimostrava havere grandissima potentia el quale tucto di bellis-
simi drappi | era vestito et uno baculo teneva nella mano dextra et una palla 122r
di puro oro nella summità di quello existeva. Et le loro sedie erano da quattro griffoni rapaci per l'aria portate e' quali alla mia barchetta factosi vicini, quella famosa donna a nome chiamandomi in tale modo parlò: "leva dal tuo core omni paura ché io sono paratissima scamparti di omni periculo et di fare cessare la impetuosa fortuna et guidare la tua navicella in quieto porto". Alle cui parolle l'inpetuosi venti, deposto loro ferocità, cessarono di spingere le motive aque et li obscurissimi nuvoli dettero luogho allo luminante Apollo in modo che con li sui fulgenti raggi ricuperai la perduta luce et per la immensa letitia di esser evaso di tale eminente periculo, in fra me medesimo confuso non sapevo che fare mi dovesse, solum desideravo intendere qual fusse quelli dua gratiosi spiriti quali me havevono dato tanto confortativo suffragio. Et io, desidero di sapere, dimandai quella spetiosa donna mi facesse digno de intendere et havere noti-
tia di suo famoso et digno nome. Allora io senti una risonante voce | et donde si 122v
uscissi non lo potetti udire, la qual così disse con acomodatissime et melliflue parole: "Progenie è questa del famoso Lauro. El nome è quello di quella gran romana che stimò più l'honore che gemma o auro". Seguì poi quella venusta donna, dicendomi che non havessi in me timore alcuno et con la mano monstrandomi quello integerrimo homo, dissemi: "Questo mio unico amico sarà sempre ad omni tua difesa paratissimo et mai ti mancherà di qualunque favore

---

4 li dua terti] F li 1/3 10 et nella sinistra mano] F et nella destra mano 11 una palla di tucto purissimo oro] F una palla tucta di purissimo oro | et nella destra] F et nella sinistra 13 grave] K di grave 26 di sapere] F più sapere 28–29 donde si uscissi] F donde si venissi 29 potetti] K petetti

et oportuno auxilio di modo che tuta sarà sempre la tua piccola barchetta nelle profundissime aque".

Quello terribile homo dalla mia nave levatosi via disparve, ne so in qual parte si transferissi. Io, alquanto in me medesimo ritornato et con le mani giuncte genuflexo o quanto meglio potetti et seppi, li rengratiai di tale loro intrinsecha humanità quale verso di me dimostrorno et a quelli dua beatissimi spiriti infinite volte racomandami loro humile et fido servitore, offerendomi
123r sempre parattissimo et ad omni loro minima | voluntà deditissimo. Et mentre più oltre volevo dire, subito la candidissima nube disparve insiemi con loro. Et io, volendo raseptare in parte la vexata navicella, di vele et remi exercitato, dal sonno mi sveghiai tucto attonito et intorno guardando se ero nel mare o dove, trovami essere nella mia cameretta con la mano in su del libro quasi parato a scrivere. Onde suspeso et pensoso stando, mi rimembrai a pocho a pocho questa veduta visione et deliberai in questa mia operetta a scriverla con quella brevità è suto possibile. Al presente intendo seguitare la terzia parte del comminciato libretto nel quale anoteremo qualche piacevole caso mathematicho. Et in questa tertia parte diremo: primo caso, secondo caso, etc.

123r **Primo caso di uno corvo che voleva bere in canpagna**

Havemo detto di sopra quanta sia la forza et virtù del numero nel fare varii et diversi giuochi, comme nelle due parte superiori si è mostro. Hora mostreremo in questa tertia parte qualche caso piacevole, e' quali con el braccio della
123v mathemati|cha disciplina se absolvano et parte variaremo la materia ché sempre tractando del multiplichare, partire et racorre saria fastidioso. Et perhò, in questa parte andaremo variando li casi che da noi saranno proposti et mostreremo quanta sia la forza et virtù mathematicha che in omni caso si extende nel parlare. Anchora si può dire di belli dubii, li quali, prestandoci adiuto lo altissimo Redemptore, legiendo, tucti troverrai. Et in questa parte diremo che questo sia primo caso, etc.

Advenne uno giorno che essendo in canpagna di Roma dove è grandissima charestia di aque, essendo forse 30 metitori quali metevano in uno spatioso piano et, comme è usanza nelli luoghi dove è charestia di aqua, la mattina quando vanno al loro exercitio, si fornisce ciaschuno di quella quantità di aqua che credano lograre in tucto quello giorno et portano in certe brocche che nella loro boccha sono strecte, et poi si stendono giù largamente et sono

8 deditissimo] F desideroso 10 di vele et remi] K di vele remi 15 seguitare] F seguire 18 voleva] K volere 20 diversi] F diferenti 20–21 mostreremo] K mostremo 24–25 mostreremo] K mostremo 25 quanta] F quanto 33 credano] F credono | lograre] F logolorare

di grandissima capacità. Hora, comme è dicto, ritrovandosi li prefati metitori nel luogho prefato | et havendo passato li dua terzi del giorno, erano tucte loro 124r
brocche vote et solo una vi restava, la quale era piena uno tertio et li dua tertii vota. Et era puntellata et rinchalzata in modo che movere non si poteva se prima non fussino levati i puntelli che dintorno vi erano. Donde, fortuna forte, essendosi scostati li metitori per alquanto spatio da epsa brocca, la quale insiemi con loro bazichature da parte tenevano, vi capitò uno corvo sitibundo che per la charestia dell'aqua cerchava cavarsi la sete in qualche modo. Et per adventura capitò dove li metitori havevano la sopradicta brocha relapsata et cerchava a quella satiare el suo apetito. Et visto che lo suo beccho all'aqua non poteva agiugnere, cercava la sopradicta broccha rompere, il che inpossibile li era. Et visto non poterli riuscire, cerchò dall'altra parte farla pendere et versare. Et visto tucti li modi essere scharsi et inpossibili, quasi comme disperato da tale inpresa levatosi, cerchando se altro ne potessi trovare che a suo proposito fusse più comodo. Et cerchato tucto quello paese non vi tro|vava nulla, di 124v
modo che più volte tentò se quella broccha potessi in modo alcuno aterrare, il che fare indarno si operava. Onde tucto pensoso et fantastico immaginava se modo o via vi fusse che a fare suo disio contento trovare potessi. Si domanda in questo caso che modo terrà questo corvo a bevere senza aiuto nessuno et non movendo decta broccha. Ché veramente questo corvo usò uno puncto mathematicho et fece in questo modo: prese con suo beccho tanti sassolini et misseli in questa broccha che l'acqua, la quale naturalmente sempre sta sopra li sassi, venne alla sommità di dicta broccha in modo che lui facilemente poteva la sitebunda sua voglia adempiere. Et benché li metitori del facto si acorgessino, per vedere tale effecto pigliandosene piacere li lasciorno la sete thraere, siché potiamo dire el corvo predicto essere mathematicho.

Secundo caso di tre mariti et tre moglie gelose et quattro et cinque

Voglio mostrare modo a ffare quella favola di tre mariti et tre moglie gelosi per chi non | lo sapessi et anchora, per chi lo saprà, dirò qualche cosa in questa 125r
di piacere. Capitando adunque 3 mariti con loro donne secho ad uno fiume grandissimo, el quale senza la barcha era inpossibile potersi dalla altra banda transferirsi. Et erano in paese costoro che non senza qualche suspecto si caminava et maxime nelli nocturni tempi et era luogho silvestre dove era charestia di habitationi. Et perché havevono loro giornate ordinatamente conpartite, non bisognava che troppo badassino per non essere alla canpagna scoperta dalla

5–6 fortuna forte] F forte fortuna 8 charestia] K charestiva 12 dall'altra] K dalla l'altra 14 altro] F altrove 29 lo] F la

obscura nocte sopragiuncti. Onde, giuncti a questo fiume et desiderosi passare via con velocità, non vi trovorno el barcharuolo che subito li haveria passati, ma solo trovorno una piccola barchetta alla ripa ligata. Et quella presa et dalla ripa disciolta, et volendo in epsa tucti ad uno tracto passare, viddono per experientia che dicta barcha non era capace di più che dua persone per volta. Et così quasi confusi rimasono. Et desiderosi passare via, pensavono el modo pas-
125v sare senza gelosia et suspecto l'uno dell'altro. Si domanda in questo ca|so se è
possibile farlo. A che si risponde che sì, et fia in questo modo: prima dirai che passi in là due donne et poi una rimangha di là et l'altra rimeni la barcha in qua. Et meni de là l'altra donna et sono in due volte passate tucte tre le donne di là. Et li homini sono di qua et le donne hanno secho la barcha, dunque le donne si potevono andare con Dio et lassare abaiare li loro mariti de là dal fiume, ma perchè omni una era gelosa del suo marito, non harieno per niente facto mai tale errore. Onde per tornare a nostro proposito et volendo fornire di passarli, farai hora passare una delle 3 donne allo in qua et così el suo marito. Et li dua mariti di quelle che sono di là dal fiume montino in barcha et vadino a trovare le loro donne. Et hanno la barcha seco li dui mariti con le loro donne, et resta a passare uno marito et una moglie. Hora passi allo in quà uno marito con la sua donna et haremo di qua dua mariti con le donne loro et di là sarà uno marito
126r et la donna soli et la barcha è di qua. Hor passino di là li | dui mariti lassando
le loro 2 donne di qua et poi mandino con la barcha le donne che è già passata per una delle dua che restano a passare. Et poi mandino o el suo marito opure una altra donna per la tertia, ma più conveniente pare a mandare el suo marito. Et così le harai passate tucte senza gelosia o suspecto l'uno dell'altro. Non che habbino ad essere traditori l'uno dello altro, ma per levare via ruggine tra loro intrinsecha amicitia, perché el suspecto che ha uno della sua sposa o donna è inmenso et grandissimo et non si può armare mai el suspecto, siché questo modo sopradicto leva via omni fantasia. Potrai tu poi col tuo ingenio fare di 4 mariti et 4 moglie, di 5 et di 6, etc. Ma nota che la barcha bisogna che sia capace di uno mancho che il numero delle coppie che sono, perché se sono 4 mariti, tante moglie, bisogna che la barcha sia capace di tre persone, et se fussino 5, bisogna che la barcha sia capace di 4 persone, et così in infinito. Et altrimenti indarno ti affaticharesti. Et perhò, quando tu proponi che sieno 4
126v et la barcha | sia capace di 2 persone et non più, fia inpossibile et farai fanta-
stichare le brighate et potrai mettere pegno, et così se fussino 5 et la barcha fussi capace di 2 o 3 al piú, similmente saria inpossibile.

---

14 a nostro] F al nostro 25 dello altro] F al'altro

Caso tertio di uno partito di monete

9	9	9	9
7	7	7	7
5	5	5	5
3	3	3	3
1	1	1	1

Fassi uno partito con quarteruoli, grossi o quattrini o ducati o quello voli, in questo modo: poni in tavola uno monte di 100, dividili in 20 monticelli che 4 ne sia 9 per monte, et altri 4 ne sia 7, et altri 4 ne sia 5, et altri 4 ne sia 3, et altri 4 ne sia uno per monte. Et in tucto harai 20 monti et in tucto sono 100 grossi o che si sieno non inporta. Hora dirai alta voce a tucti: "chi di voi con 5 di questi monti fa 30 apunto, guadagni tucti questi grossi et se fussino ducati più ghiotta vi sara la brighata". Et nota che quando lui ha preso una posta che quella non possi mutare et dire: "io mi voglio repigliare". Et chi non fa 30, dirai habbi a pagare uno tanto, verbi gratia, una cena o uno grosso o uno bolognino. Et quanto più piccola sarà la pena | tanto più gente vi si metterà a farlo. Questo caso dico 127r essere inpossibile che con queste poste pigliandone 5 che tu mai possi fare 30 apuncto. Et per tanto mecti el partito secretamente et sicuramente ché niuno numero caffo, comme altra volta ti ho decto, multiplicato per uno numero caffo mai può produrre numero pari. Questo vedrai tu provandolo da te a te. Sarà bello, et quando la brighata ti dicessi: "fallo tu questo", et tu dirai non lo volere fare perché lo inpararebbono.

Caso quarto di uno altro partito di monete

10	10	10	10
8	8	8	8
6	6	6	6
4	4	4	4
2	2	2	2

2 voli] F vuoi   5 grossi] F o grossi

Fassi uno altro partito simile al precedente, ma nelle poste contrarie. Et porrai pure 20 poste delle quali 4 ne sia di 2 per posta et 4 ne sia di 4 per posta, et 4 ne sia di 6 per posta, et 4 di 8 et 4 di 10 per posta, et harai 20 poste et in tucto harai 120 ducati in tavola. Hora dirai: "quello che con 6 monti farà 35 apuncto, guadagnerà tucti questi ducati". Non lo faccendo, paghi tanto per una cena comme a te pare. Anchora dirai questo caso essere inpossibile ché el numero pari multiplicato per el numero pari non può mai produrre se non numero pari et perhò 127v è inpossib|ile, onde tu potrai securamente porre questo partito, ma per più mostrare di havere paura di non perdere, dirai che quando uno ha preso una posta che non si possa ridire et dire: "io voglio mutarmi" et pigliare una altra, comme tu saprai con una certa gratia proporre alle brighate el partito.

Caso quinto d'uno che fa 3 domande a uno tracto

Intervenne che una volta fu proposto uno caso in questa forma. Uno prese certa quantità in mano di danari di che sorte vuoi, o quattrini, o ducati, non inporta, et disse a uno altro in questo modo: "dammi tanti grossi quanti io ho in mano et poi mi finisci con questi che io ho in mano, insino che io ne habbi 80 et poi me ne dona 10". Si dimanda per questa proposta comme lui farà a satisfarla in uno tracto. Dirò brevemente che facci così che sommi tu tucta quella quantità che lui vole che tu li finisca et con quello che ti dice volere che tu li doni et quella fia la quantità che a tucti satisfarrà. Ecco lo exemplo: pognamo che lui havessi in mano 50 et ducati ne voglia tanti quanti lui ha in mano che sono 50. Et poi vole 128r che tu li fi|nischa insino in 80, che a finire 80 li mancha 30 et hai a pigliare 30 et 50 di prima, fa 80. Di poi te dimanda volere che tu li doni 10, harai 90. Dunque per volere satisfare a sua domanda ti bisogna pigliare 90. Et se bene notasti lui ti chiede che tu li finisse insino in 80 et poi li donassi 10 che in tucto fa 90. Ecco che la regola ti servirà benissimo et sempre.

Caso sexto di uno signore che manda in uno giardino a corre rose con certe conditioni

Uno signore manda uno suo servo in uno giardino et sì li dici: "va, portami una rosa bella et gallante". Colui va et entrando alla prima porta el portinaro li dice: "fratello, sappi che di tucto quello che in questo giardino cogliessi, tu debbi a questa prima porta lassare la metà di tucto quello che tu havessi et uno più. Costui, ulterius procedendo, andò più avanti alla seconda porta, dove li fu dicto dal portinaro che usanza era qualunche cogliessi pomi o arancie o limoncini o rose che la metà et uno più a quella porta doveva lassare. Et così li fu decto

---

10 io voglio mutarmi] F io mi voglio mutare 22 et hai a pigliare 30] F *omissis* 26 benissimo et sempre] F sempre et benissimo 29 si li dici] F si li dicine

alla tertia porta che quello giardino non haveva più che tre | porte. El signore 128v
li comandò che non ne portassi se non una. Dunque, a costui bisognerà adoperare la doctrina mathematicha, perché se per ventura ne havesse colte più una che 'l bisogno, el signore stimava tanto queste rose che veramente lo haveria cacciato via. Onde pensando da sé, vidde che li ne bisognava corre 22 ché venuto alla prima porta ne lassò la metà che sono 11 et una più, che sono 12, et a llui ne rimase 10. Et venuto alla seconda porta, ne lassò 5 per la metà et una più, fa 6, restaline 4. Et venuto alla tertia porta lasciò 2 per la metà et una più, fa 3, et restaline una francha. Et quando fussino 5 o 7 porte, terrai sempre questa regola generale che ti servirà a quante porte tu vuoi. Farai così: comincia alla prima porta del giardino quando tu entri dentro et dirai: "io ne voglio una francha". Et a questo poni una, fa 2, multiplichalo overo, diciamo, doppialo, fa 4, et tante bisogna che tu ne habbi franche quando tu vieni alla prima porta. Hora per la
seconda, per regola generale, poni sempre uno più et poi adoppia el | tucto per 129r
haverne 4 franche. Alla seconda porta poni uno in su 4, fa 5, et adoppia 5, fa 10, et questo bisogna sia francho alla seconda porta. Et per volerne 10 franche porrai uno in su 10, fa 11, quale doppiato fa 22. Et questo ti bisogna corre nel giardino. Ma se tu havessi a passare 5 porte, vedrai per la data regola che a te bisogna corre 94 rose per haverne una francha. Et se tu ne volessi 3 franche o 4 o 6, etc., per la regola data tucti li simili casi si potranno absolvere facilmente et con prestezza.

Caso VIIo di uno signore che fa guardare una ciptà

Intervenne che uno signore haveva una sua ciptà, la quale dalle partialità intrinseche era molto tribulata. Et comme in simile ciptà interviene che una volta più che l'altra sono più periculose. Uno giorno per uno suo secretario li fu rivelato che se non rimediava presto che loro erano per tagliarsi tucti a pezzi. Onde el signore, comme homo generoso et di suo popolo amorevole, ordinò subito bono provedimento et pensò che il guardare le porte fusse cosa salu-
tifera che fare vi potessi. Et perché | la predicta ciptà haveva apuncto 8 porte, 129v
ordinò delle 8 guardarne 7 et una lassarne libera che omni uno potessi entrare et uscire liberamente. Et a questo fare ordinò 7 valenti et famosi conestabili, in simile exercitio molto pratichi et experti. Et dato loro certa quantità di fanti, disse loro che con più velocità che possibile fusse, alla prenominata ciptà si rapresentassino, il che molto desiderava, et che senza tumulto o rumore alcuno entrassino. Haveva questa ciptà, comme qui in margine vedi disegnato, ad omni porta 2 strade, le quali andavano a dricto verso delle 2 porte a riscontro, siché

4–5 haveria] F harebbe 9 porte] K parte 27 onde] F donde 31 valenti et famosi] F famosi et valenti | conestabili] K contestabili 32–33 disse loro che] F disse che

entrando per una porta potevi pigliare la via andare verso 2 porte et ad andare a una porta per 2 porte potevi entrare. El signore comandò loro tenessino questo ordine che non si fermassino a quella prima porta dove entravano, ma che fingessino passare via, andassino adirittura a una delle 2 porte quali volessero et quella piglassino et lì stessino. Et comandò loro che, sotto pena della testa,
130r comme havevono preso una porta non vi lassassino entrare | né uscire nessuna persona se non chi havessi el salvoconducto o bulectino di mano propria del signore. Et così anchora chi di loro si fermassi alla porta che lui giugneva di prima, facevali mettere di pena el capo per la sopradicta legge. Costoro hanno a considerare che comme una porta da uno di questi conestabili è presa, non vi può più entrare l'altro conestabile per andare all'altra porta. Donde bisogna, avanti che al luogho pervenghino, habbino ordinato et pensato el modo che tenere debbono in fare lo effecto loro et non preterire el precepto del signore loro. Si domanda in questo caso se è possibile che loro possino farlo. Il modo dassi a questo caso è il seguente, et così a tucti e' simili a questo. La ciptà, comme è dicto, ha 8 porte et sono 8 littere comme vedi: A, B, C, D, E, F, G, H. Dico dunque che a volere tenere questo ordine sempre, el conestabile che viene occupi quella porta donde è entrato quello che innanzi a llui. Comme se 'l primo che viene, entri per la porta A occupa la porta F, el secondo entri per la porta D et occupi la porta A, il terzo entri per la porta G et occupi la porta D, et poi el quarto entri per la porta B et occupi la porta

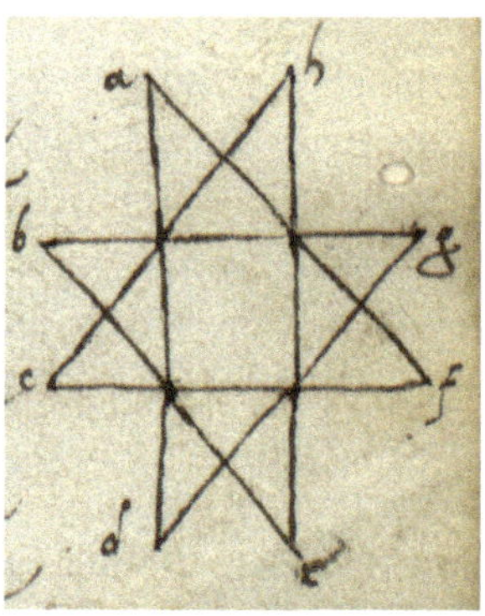

FIGURE 5
Drawing on folio 130r

130v G, el quinto entri per la porta E et occupi la porta B, el sexto entri per la porta H et occupi la porta E, el septimo entri per la porta C et occupi la porta H. Ecco che tu hai servato l'ordine del signore et hai preso 7 porte et solo una, cioè el

---

4 volessero] F volessino 7 salvoconducto] K salvuoconducto 9 mettere di pena] K mettere pena 15 dassi] F darassi 19 per la] F nella 22 G] F C 23 E] F *omissis*

C, rimane libera. Et per questa vole el signore che li homini della terra possino entrare et uscire per loro facende. Siché potrai fare per questa regola di più sorte a tuo modo simili a questa.

Caso VIIIo di uno che manda tre sua famigli in merchato

Uno homo dabene, faciendo in Lione una grossissima ragione et havendo bisogno di multitudine di ministri et volendo havere homini experti, quando li erano messi innanzi dua o 3 giovani che li tenessi alla sua ragione, in questo modo li usava a sperimentare. Dava a uno 10 peze di saia che le vagliono pocho et al'altro ne dava 20 et al'altro 30 et diceva loro: "andate et vendete queste saie tucte a uno medesimo priegio et al fine mi portate tanti denari l'uno quanto l'altro". Si domanda in questo caso el modo che tenevano li sopradicti. A che rispondo che in su la prima giuncte, trovando essere dovitia | 131r di simile merchantia, ne dettono una al ducato. Colui che ne haveva 10, ne vendé 7, et parendoli gittarle via, non ne volse vendere più. Poi questo medesimo merchatante comperó dalli altri dua al medesimo priegio. Quello delle 20 ne dette 18 et il resto tennono forse per non uscire di commissione et mostrare al loro maestro che erano mathematici. Et forse quelle che lor o serbavano erano più belle che le altre et speravono di haverne al fine della fiera più. Hor sia quello si vole, piacque loro fare così. Et il terzo ne vendette delle sue 29 et restonne una. Dunque el primo ha 7 ducati et 3 peze, el secondo ha diciotto ducati et 2 peze, el tertio ha 29 ducati et una peza. Hora, essendo quasi finita la fiera, non vi era rimasto più saie che queste. Et uno volendole, bisogno che la comperassi undici ducati d'oro l'una. Dove colui che haveva 7 ducati et 3 peze fece 33 ducati et 7 ne haveva prima, fanno 40. Et quello che haveva 18 ducati et 2 peze fece 22 ducati et 18 ne haveva, fa 40. El tertio, che haveva 29 ducati et una peza, ne fece 11, 29 ne haveva prima, fa 40. | 131v Eccho che tucti venderno a uno priegio et tucti portorno al maestro loro ducati 40. Onde el maestro, atteso che questi erano boni ragionieri et boni matematici, li acceptò nella sua ragione. Et di questi casi infiniti se ne può porre, li quali tucti potrai, servando questo ordine, facilmente absolvere, che si pone casi di 4 et di 5 che tucti vendino a uno priegio et alla fine portino uno medesimo danaio. Et per questa sola regola li potrai solvere tamen a satisfactione di quelli non intendono. Ne porremo uno di 4 medesimamente per experimentarli.

Caso VIIII di uno che manda 4 garzoni a una fiera

Mandò questo medesimo mercatante 4 garzoni a una fiera et a uno dette 10 peze, al'altro 30, all'altro 50, al'altro 70. Et disse loro che non tornassino a llui

---

22 uno] K una 27 questi] F loro 32 quelli non] F quelli che non 35 mandò] K dando

se prima non havessino vendute tucte et che vendessino comme li superiori, a pari priegio et che tanti danari portassi l'uno quanto l'altro. In questo caso si dimanda el modo che loro tennono, quali sono varii et innumerabili. Ser-
132r vando le valute et la quantità delle cose, poterno costoro in su la prima | giuncta vendere dicte peze di saia et darne 7 al ducato che el primo fece uno ducato et avanzolli 3 peze, et el secondo ne fece 4 et avanzolli 2 peze, il tertio ne fece 7 ducati et avanzolli una, el quarto fece 10 ducati apunto et non li restò nulla. Hora, in su la sera alquanto rincarorno et quelle che li restorno si venderno 3 ducati la peza. Dunque el primo che haveva uno ducato delle sue 3 peze ne fece 9 ducati, et uno ne haveva, fa 10. Et quello che ne haveva 2, fece 6 ducati, et 4 ne haveva, fa 10. Et quello che ne haveva uno ne fece 3 ducati et 7 ne haveva, che fa 10. El quarto le vendé tucte. Dunque, dirai che portorno 10 ducati per uno. Possi fare per altro modo et fia questo. Se 'l patrone havessi dato al primo 10 pezze, al secondo 20, al tertio 30, al quarto 40, con la medesima conditione, dirai ch'el primo ne dette 7 per 7 ducati et avanzolli 3, el secondo ne dette 18 per 18 ducati et avanzolli 2, el tertio ne vendé per 29 ducati 29 peze et avanzolline una, el quarto ne vendé 40 per 40 ducati et non avanzò nulla. La sera poi rincharorno le saie. Venderno loro quelle, li restorno ducati 11 l'una. Onde quello
132v che haveva | 7 ducati li restò 3 saie et fece 33 ducati et 7 ne haveva, fa 40. El secondo che haveva 18 ducati et 2 peze, fece 22 ducati, 18 ne haveva, fa 40. El tertio haveva 29 ducati et una peza, fece 11 ducati et fa 40. El quarto anche haveva 40 ducati. Et così portorno 40 ducati et venderno tucti a uno medesimo priegio. Et con questi dua sopradicti modi potrai tucti li simili casi solvere quando ti fussino proposti.

Caso 10 di una mela divisa in uno capello

Fassi anchora uno altro partito che si toglie una mela et mettisi in uno capello, quando voi fussi a una tavola 10 o 12 persone o 20 o 25, quanti più sono meglio. Et tu, senza dire oniuno quello che tu intendi di fare, piglia la mela et fanne 2 parte et poi rimetti nel capello et dallo a quello che ti sia da canto et dirai che pigli uno di quelli pezi et faccine 2. Et poi di mano in mano seguita per infine, ché tucti ciaschuno habbi facto el suo pezo. Et quando tucti hanno fornito et tu dirai alta voce: "quello che indovina quanti pezi sono questi che sono qui den-
133r tro, habbi a dare una parmata per uno". | Overo dirai: "quello che non indovinerà pagherà uno bolognino per una cena". Et in quello modo che a te parrà di dire farai tu. Et quando non havessi mela, tolli uno pezo di carota et omni cosa frangibile serve. Si domanda quanti sono li pezi della mela o carota. A tuo modo

---

3 dimanda] F domanda 4 le] K la 16 avanzòlline] F avanzòlli 33 indovinerà] F indovina

dirai che li pezi sono uno più che non sono li homini che sia la verità. Anovera et vedralo tu. Pigli una mela. Dirai dunque "uno" et poi tu fa uno pezo più che sono 2 et l'altro, pigliando uno delli dua pezi farà uno altro che siranno 3 et l'altro di mano in mano, in modo che sempre alla fine saranno uno più che li homini che sono a cerchio. Et questo è colpo matematicho. Et similmente se tu dicessi "egli è uno stidione di becchafichi 20, che omni becchaficho è nel mezo di dua lardelli. Dimando quanti lardelli bisogna". Dirai che bisogna uno più che non sono li becchafichi et bisogna a 20 becchafichi 21 lardelli. Et così in simili casi sempre risponderai. Ma lo ignorante crederà che ad omni becchaficho bisogni 2 lardelli et dirà che 20 becchafichi vogliono 40 lardelli et tu dirai che non è la verità | et rimarrà confuso. 133v

Caso XIo di levare 100 sassi a filo o caminare uno miglio

Usasi proporre uno partito in questa forma, cioè dire a uno: "o veramente tu vogli caminare uno miglio o tu vuoli levare 100 sassi o petre che fussino a filo posti l'uno distante dal'altro uno braccio". Et in questo partito bisogna essere buono mathematicho perché chi non è bene experto, in questo caso abagliarebbe. Dimando adunque qual sia meglio, o pigliare a caminare uno miglio intero overo di levare li cento sassi? In questo bisogna sapere uno miglio quanto è, che nella mia ciptà si fa uno miglio giusto braccia 2000. Hora, saputo questo tu dirai: "io ho a racorre 100 sassa a filo et tucti racorli in uno cerchio o in uno cestone o paniera, comme si sia, overo caminare 2000 bracci". Hora a te bisogna racorre tucti li numeri che sono inclusive di uno, fine in 100. Et racolti, con velocità troverai che sono 10100. Faralo in questo modo: per racolli presto, multiplicha la distantia delli passi che sono 100, multiplicha 100 via 100, fa 10000 et sopra questo poni | el numero delli sassi che sono 100, farà 10100. 134r
Et tanti sono li passi che ti bisogna fare a llevare 100 sassi. Siché vedi che vantaggio è che più presto vorrei caminare 4 miglia che racorre li 100 sassi. Perhò a simili casi ti bisogna havere el capo a butegha et non te ne lassare menare, ma quando tu facessi a vista o vuoi andare sino a Careggi et tornare o vuoi levare 100 sassi a filo posti distanti l'uno dal'altro uno braccio, bisogna sapere quante miglie giuste sono dal luogho ove tu se' a levare e' sassi a Careggi et fare quello conto et poi pigliare quello dove vedi sia el tuo vantaggio. Et così, quando fussi messo uno partito di uno luogho che a te non fussi noto, bisogna havere grande advertentia a questo, a non volere esserne menato per il naso ché chi non è praticho vi è dentro grandissimi errori, dunque starai in te.

---

2 fa] F fai 3 siranno] F saranno 4 alla fine] F al fine 8 lardelli] K lardello 21 bracci] F omissis 23 10100] F, K 101001 25 10000] F, K 1000 | sopra questo poni] F sopra questo sempre 35 vi] K vie

Caso XII di octo donne che fanno uno ballo et septe giovani

Ritrovandomi una volta alla festa che nella quatragesimale domeniche si fanno
134v nel nostro | Fesulano monte et essendo andato in conpagnia di certi altri gio-
vani a me eguali, eravamo per numero 7 tucti. Et udito el vespero, comme
si fa chi va qua et chi va là, cerchando qualche solitario luogho per potere
alquanto refrigerarsi con bere o mangiare qualche cosa, a tale refrigerio dipu-
tata, advenne per caso che noi, usciti dello divoto templo et presa la debita
indulgentia, voltammo su per uno suavissimo colle di veridissimi arburcelli
ornato. Et dalla suavità del loco menati, ci transferimo in quello per spatio di 2/3
di miglio senza mai alcuna persona trovare, desiderosi vedere che fine havessi
lo amenissimo viaggio. Alla fine sua pervenimmo in uno vagho praticello di
vaghi et innumerabili fiori ornato. Non era di troppa grandezza, ma intorno
intorno dalli arburcelli circumdato in modo che li raggi dello aurato Apollo
pocho lo potevano offendere, solo per certi parvulissimi spatii fra ramo et ramo
in quello si vedeva penetrare. Giuncti adunque et pervenuti in quello spetioso
luogho, vedemmo in epso permanere 8 pulcherrime et spetiose giovane, le
135r quali, in primo aspecto, | tucti insiemi dubitammo quelle essere nimphe alla
pudicissima Diana consecrate, ma le aurate veste ciò non essere ne dettono a
noi lume, perché tali Nimphe simili hornamenti non usano, ma bianchissime
veste et purissimi habiti. Certificati adunque quelle essere donne a noi equale,
a quelle factoci più propinqui, vedemmo quelle essere per numero 8. Et pre-
sesi una l'altra per mano, cerchavono quello felicissimo giorno in grandissimo
dilecto consumare. Et erano tucte di sì mirabile belleza hornate, che stupidi
et attoniti a contemplare loro signorile aspecto alquanto ci fermammo. Donde
da uno delli nostri conpagni furno le dicte giovane domandate se noi nel loro
festivo ballo acceptare volevano ché se non li era molesto, volentieri con loro a
danzare ci fermeremo. Alle quali parole una di epse che nello aspecto di tucte
le altre era guida et capo, così rispose: "noi siamo tucte contente che voi entrate
nel nostro ballo, con questo pacto che uno di voi cominci a contare da una di
135v noi quale più vi piace et da quella a che mano vole vadi contando sino | a 4
et quivi si fermi et con quella si acompagni". Et così seguino di mano in mano
con legge che comme una di noi è acompagnata, da quella più non possiate
cominciare a contare, ma bene si possino le acompagnate mettere nel numero
delle 4, contando et non servando questa legge, insino che tucti vi acompa-
gnate. Al tucto siate privi di potervi insiemi con noi acompagnare, essendo voi 7

---

6 bere o mangiare] F mangiare o bere 17 aspecto] K aspecti | quelle] F *omissis* 19 tali Nimphe simili hornamenti non usano] F tali non pigliano simili hornamenti né usano 21 vedemmo F] dicemmo

et noi 8, una sola resterà guida et capo del ballo. Dunque piglierete modo di fare tale effecto overo nella quiete che ci trovasti, in quella ci lasserete alla nostra consolatione. Seguendo noi tale proposta intesa, desiderosi con loro acompagnarci per l'amirabile et infinita belleza che nelli loro delicatissimi volti si dimostrava, onde standо alquanto suspesi, trovammo modo di fare tale opera et in questo caso operamo la scientia mathematicha. Se domanda l'ordine di fare tale effecto quale fia questo immediate seguente. Sia el cerchio tondo, comme qui vedi in margine. Et li tondi che vedi in decto cerchio sono le 8 donne et omni una ha la sua lettera dell'alfabeto comme vedi: A, B, C, D, E, F, G, H. Hora

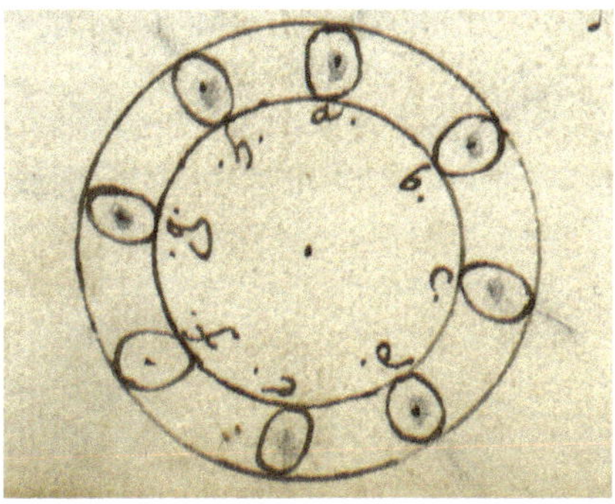

FIGURE 6
Drawing on folio 135v

comme è dicto, bisogna cominciare da una et seguire, numerando sino in 4. 136r
Che diciamo si cominci dalla A et vadi a mano dextra, et la prima che si acompagnerà sarà lo F. Et quelle che sono acompagnate hanno el puncto nel mezo, comme vedi. Hora a volere acompagnare tucte le altre, per regola generale commincerai in luogho che tu acompagni quella dove ti comminciasti a noverare comme qui comminciasti dalla A et acompagnasti F. Hora ti comincia in luogho che tu acompagni la A et cominciati dal D et acompagnerai la A. Al presente cercha acompagnare el D che cominciandoti dal G anovera sino a 4, acompagnerai el D, poi comincierai dal B et acompagnerai el G. Et poi ti comincerai dal E, acompagnerai el B, poi commincerai dalla H et acompagnerai la E. Et poi comincerai dal C et acompagnerai la H et solum ti resterà el C che non sarà acompagnato. Et noi per acompagnarci con queste giovane tenemmo questo modo. Et omni uno con questa regola si acompagnò con quella che più li piaque. Et così passò quello | giorno in suavissimi canti et balli, et poi leti ne tornammo 136v

6 operamo] F operare 9 dell'alfabeto] F dello alfabeto 10 è dicto] F dicto è 17 G] F *omissis* 18 comincierai] F ti cominci | G] F *omissis* 19 E] F *omissis* | E] F *omissis* 20 C] F *omissis* | solum] F solo | C] F omissis 23 suavissimi] F suassimi

omni uno al suo habitaculo. Et perché questo caso in sè ha parte di mathematicha, l'ò voluto in questi nostri casi inserire et con questo calculando potrai venire alla notitia di molti altri casi di più et di meno, secondo che tu vorrai.

Caso XIIIo di uno che compera et vende a uno medesimo priegio et guadagna

Essendo uno giorno a uno cerchio a ragionare, comme spesso adviene, uno buono abachista mi disse uno caso che molto mi fece fantastichare. Tamen lo ritrovai et parmi degno di essere connumerato nelli mathematici ludi. Et è questo: uno compera certa quantità di ova et poi li rende a uno suo amico per el medesimo priegio che li comperò et guadagna. Si domanda el modo che lui fece. Dirai che fece così che li 100 ova li costassino soldi 40 che se tu fai bon conto, li ne vene 2 1/2 al soldo. Havendo costui comperate le ova, essendo per andarsene, li sopravenne uno suo carissimo amicho et preghòllo
137[r] che li rendesse | li decti ova. Lui disse essere contento che li ne renderebbe una parte che farebbono da buono compagno che li partirebbe a mezo con colui, et numerando li ova trova apunto haverne 200, onde li disse: "se te ne vuoi 100, io te li darò et di 50 te ne voglio dare 3 al soldo et delle altre 50 te ne voglio dare 2 al soldo ché così le ho comperate io. Che a questa ragione te viene 5 per 2 soldi et così costono a me". Hora io ti ho dicto di sopra che el numero delli ova li costò soldi 40 et volli vendere al suo amico et darline 50 a 3 per uno soldo che li 50 ova montano soldi 16, danari 8 et li altri 50 ova li dà per 2 al soldo che montano 25 soldi. Hora acoza 25 soldi et 16 soldi et 8 danari, farà 41 soldi et 8 danari. Ecco che costui rivende per il medesimo priegio che comperò et guadagna soldi 1 danari 8. Et questo è la mirabil forza del numero per la varietà del vendere. Possi anchor fare per una altra via questo medesimo effecto. Comme uno trova uno altro suo amico, dicendoli: "io vorrei comperare certa quantità di ova et, cerchando, non le trovo". Onde l'amico suo si offerì servìllo et disseli: io ne ho
137[v] comperate | per me et per servirti, te li renderò mezi et li altri voglio per me, ma non voglio perdere. Et vendélli per quello li comperò, guadagnò et numerando li sua ova vidde che ne haveva comperato 2 mancho di 200. Hora lui disse: "io te ne renderò la metà" che sono 99 et a llui costarono li 75, ne hebbe 3 al soldo che li 75 li costorno 25 soldi et li 24 li costorno più che li comperò, soldi 16, ché ne hebbe uno et mezo al soldo che in tucto li costorno li 99 ova soldi 41. Hora lui disse: "tu mi darai di 72 24 soldi che così costorno a me, et di 27 mi darai 18 soldi che tanto spesi et in tucto li dette 24 et 18, fa 42. Ecco che lui guadagnò uno soldo et vendé per li medesimi priegi che comprò. Et questi dua modi bastino che molti più ne potrai da te medesimo fabbricare et similmente potresti dire che uno compera et rivende a quello medesimo pregio et perde che allo in-

14–15 et numerando] K et numerando et numerando

gnorante et quasi alla magiore parte parrà caso inpossibile, et pure è et vedi per prova che riesci.

Caso XIIIIo di dua fratelli quali hanno a dividere una botte di vino

Intervenne una volta comme si usa hoggi et maxime nelli homini ingnoranti 138r et poveri et la magior parte nelli rustici che comme sono nati vogliono partire dalli altri fratelli et quando non hanno altro che partire, partono lo amore.

Essendo adunque al Ponte a Sieve nella mia villa quale, comme omniuno sa, è apresso alla ciptà fiorentina 10 miglia, et andando comme si fa vedendo le mia parvolette posessioni, havevo in uno mio poderetto, chiamato dalli paesani "Le valli", dua fratelli e' quali uniti insiemi havono tolto quello podere da me a lavorio comme s'usa.

Et uno giorno infra li altri vennono per certa differentia a parolle molto contentiose et di odio piene, per le quali parolle uno di loro, più superbo che l'altro, disse non volere con seco più permanere, ma voleva partire et che omni uno facessi da sé li facti sua. L'altro, che irato et furibundo era, disse: "cotesto fia a me sommo piacere et avanti che di qui ci partiamo, io voglio che omni uno pigli le sue bazichature et stia sé da sé che non habbiamo tanta robba che in una meza hora non si parta". Onde d'acordo | insiemi partirno omni cosa, et 138v grano et masseritie et tucto. Solo li restava a partire una botte di vino, la quale era di tenuta di barili 8. Et venendo a partire questa botte, non havevono barili. Onde uno disse: "lassamo stare questa sino che andiamo al castello per li barili". L'altro, che dalla furia era acechato, disse: "io non sono per partirmi, che questa botte si partirà se io dovessi sfondarla". Onde sopragiunsi apuncto a questa loro differentia et provato prima se con parole amorevoli dalla loro pertinacia li potevo rimuovere, il che visto essere inpossibile, deliberai metterli d'acordo. Et inteso loro openione dissi: "io vi contenterò". Et entrato in casa loro non vi trovai niente aperto a potere dividere questa botte. Solo vi trovai una altra botticella che teneva 5 barili et uno tinello che ne teneva 3. Et benché cognoscessi el caso havere in sé qualche difficultà, per levare via li scandoli che fra loro potevano nascere, me industriai tanto che ciaschuno di loro hebbe la sua parte giustamente senza achattare altra misura. Si domanda in questo caso el modo. Et tu dirai che fece così che | la prima cosa lui empì quella tinella che teneva 3 barili 139r et poi la tinella votò in su quella botte che teneva 5 barili et nella botte grande ne resta 5 barili. Hora una altra volta con la botte grande empi quella tinella et nella botte grande ne resta 2 barili. Et poi con quello vino della tinella finisci

---

9 poderetto chiamato] K poderetto et chiamato 10 havono F] havevono 10–11 a lavorio F] a llavorare 12 infra F] fra 22 partirmi F] partirmi di qua 28–29 havere in sé F] in sé havere 33 su F] sulla

di empire quella botte che tiene 5 barili et resteranne nella tinella uno barile. Hora vota tucta quella di 5 barili in su la botte grande et harai nella botte grande 7 barili et nella tinella uno et la botte da 5 barili è vota. Hora vota quello barile che è nella tinella in su la botte delli 5 barili et poi con la botte grande empi el tinello di 3 barili et resteranne nella botte grande 4 barili per la parte di uno di loro. Et poi metti quello vino che è nella tinella in su la botte delli 5 barili dove ne è uno et 3 fa 4 per la parte dell'altro. Et così li missi d'acordo. Et con questo caso se ne può mettere infiniti, e' quali tucti per questa via li solverai. Hora nota uno altro da questo suprascripto, in tucto vario.

139v **Caso xv di 2 fratelli che similmente vogliono dividere una botte**

Uno altro caso difficile allo ingnorante si propone che sono dua fratelli che hanno a dividere una botte qual tiene 12 barili et non hanno se non 2 boticelle, et l'una tiene 7 et l'altra 5 barili. Si domanda in questo caso el modo che tennono costoro a dividere che omni uno habbi la sua parte per ciaschuno, cioè la metà. Terrai questo seguente modo: prima empirai quella che tiene 5 barili et votala in quella che tiene 7. Et poi una altra volta empi quella di 5 et con epsa finirai di empire quella tiene 7 et in quella di 5 ne resterà 3 et nella botte di 12 ne resterà 2.

Hora vota tucta quella che tiene 7 barili in su quella che tieni 12 et sarànne in su quella di 12—9 et in quella che tiene 5 ne sarà 3 et l'altra di 7 sarà vota. Hora metti quelli 3 barili che sono in quella delli 5 in quella delli 7 et poi empi quella delli 5 con quelli 9 et harai piena quella delli 5 et in quella delli 7 ne harai
140r 3 et nella | grande ne sarà 4. Finisci di empire quella delli 7 con quella delli 5, che in quella delli 7 ne hai 3 et a empierla ve ne mancha 4. Dunque cava 4 di quella di 5, ve ne resterà uno et quella delli 7 sarà piena. Et così vota li 7 barili in su la botte grande che prima ve n'era 4 et 7 fa 11 et quella delli 7 resta vota, et quella delli 5 ve n'è uno. Dunque vota quello uno nella botte delli 7 et poi con la botte delli 11 empi quella delli 5 et resterà nella botte grande 6 barili per una parte di uno di loro et per l'altro tu metterai li 5 barili che sono in quella che tiene 5 in su quella che tiene 7, che prima ve n'era uno. Et saranno 6 per la parte del secondo. Et così è soluta questa questione. Et con questo ordine ne potrai porre infiniti et varii che per questa regola et per altre se solveranno comme per te medesimo potrai provare et investighare.

10 botte] F bocte di vino 12 botte] F bocte di vino 17 finirai] F finisci | resterà] F resta 18 resterà] F resta 19 quella di 12] K quella che tieni 12 25 resterà] F resta 28 barili] F bari | una] F la 30 in su] F in sul | che tiene 7] F delli 7 31 soluta] F asoluta 33 provare] K proporre

Caso XVI di 10 homini che capitorno a una hosteria

Racontòmmi una volta uno mio amicissimo familiare di casa, el quale si chiamava Domenicho di Nicolò Dall'Antella* che una volta, rit|rovandosi in Roma, 140v
stava a una hosteria et quella governava lui, tucta la quale si chiamava l'hosteria Della Scala. Et a quella, comme è usanza, capitava molti homini da bene et merchatanti fra quali uno giorno vi capitò certi merchatanti fiorentini quali erano per numero 10. Et capitati quivi in Roma per dimorare qualche giorno per dare luogho a qualche loro merchantie, alloggiorno a questa hosteria. Et posatisi quivi comme intervene alli homini da bene, sempre loro si dilectano di motteggiare et burlare. Et uno delli altri, più astuto, dimandò questo Domenico quale era non altrimente che l'hoste in quella magione stimato, quanti ducati voleva et obligassisi a fare loro tanti pasti in quanti modi loro variatamente si potessino a tavola aseptare, cioè che comme una volta si sono aseptati a tavola, non si possino più in quello medesimo modo aseptare, che uno di loro bastava fussi tramutato. Alle quali parole, pensato alquanto el prefato Domenico et grossamente facto suo conto, disse: per 50 ducati d'oro io mi obligharei a fare simile | partito. Onde colui rispose prima al quanto stato suspeso volerli dare 141r
ducati 40 di oro se si voleva a ciò oblighare. Di che Domenico disse volere con el vero patrone conferire, che benché lui fussi ministratore et havessi piena auctorità di simili casi, senza consentimento del suo patrone non farebbe niente. Onde li dissono che andassi a dimandare el patrone prefato. Andato adunque el sopradicto Domenico et tucto conferito con el suo patrone, el quale non era più experto che il bisogno, et da sé medesimo facto conto, disse: "io non credo che li decti 10 homini si possino tramutare in più che 100 modi che saranno 100 scotti che sono tucti a uno grosso per pasto, 5 ducati d'oro. Ma diciamo che se aconcino in 200 modi che a me pare inpossibile, non siranno più che 10 ducati d'oro, che qui si guadagnerà 30 ducati. Ma quando non se ne guadagnassi più che 10, sarà buono civantio, siché a me pare che tale contracto si faccia. Et a causa che loro poi non si possino ridire, si vole fare tale obligho per mano di publico notario, perhò torna presto a ciò non si pentino et ferma questo mer|chato". Tornato adunque, Domenicho disse alli merchanti comme 141v
el suo padrone era contento et chiamato uno notario fecero e stipulorno uno contracto. Se dimanda in questo caso chi fece di loro meglio: o l'hoste o li merchanti. Bisogna, a volere tal questione solvere, fare così: vedere li 10 homini in

1 uno] K, F una 2 racontòmmi] K faccontommi (iniziale errata) 12 obligassisi] F obligarsi | variatamente] F variamente 12–13 si potessino a tavola aseptare] F a tavola si potessino asectare 16 obligharei] F obligho 22 Domenico] K Demico 32 fecero] F fecono

* Membro di una nota famiglia fiorentina, meglio non identificabile.

quanti muodi varii si possono a tavola aseptare. Et terrai questa regola generale in questo et in tucti gli altri casi a questo simili. Dirai uno non si può aseptare più che in uno modo et dua non si possino aseptare più che in 2 modi. Hora per vedere 3 in quanti modi si possono aseptare, multiplicharai 2 via 3 fa 6, cioè sempre multiplicha el numero delli homini che vengono per lo numero delli modi che hai alle mani. Comme in questo: tu hai che li 2 homini si possono aseptare in 2 muodi. Hora tu voi vedere 3 homini in quanti muodi si possono aseptare: multiplicha 3 per 2, fa 6 et dirai che li 3 homini si possono aseptare a tavola in 6 modi che sia il vero. Se el primo sarà fermo in capo di tavola, li 142r altri 2 si tramutano in dua | modi et stando el secondo in capo di tavola, li altri 2 si tramutano in 2 altri modi, fanno 4 et stando el tertio in capo di tavola, li altri 2 si tramutano, si multiplicha in 2 modi. Eccho che li tre homini si tramutano in 6 modi. Hora per vedere 4 homini in quanti muodi si tramutano, farai così: multiplicha 6 modi che si possino aconciare li 3 via 4 homini fa 4 via 6, 24 et in tanti modi, dirai, che si possono aconciare li 4 homini, che sia il vero, stando el primo fermo in capo di tavola, li altri 3 si tramuteranno in 6 modi et poi, stando el secondo in capo di tavola, li altri 3 si tramuteranno in altri 6 modi et stando poi fermo el tertio, li altri 3 si tramuteranno in altri 6 modi. Et in tucto sono 24 modi. Et hora per sapere li 5 homini in quanti modi si possono aseptare, multiplicharai 24 modi via 6 homini farà 24 via 5 fa 120. Et in tanti modi, dirai, che li 5 homini si possono aseptare in 120 modi ch'el sia la verità, stando fermo el primo in capo di tavola, di altri quattro si tramuteranno 142v fra loro in 24 modi, comme è decto, et poi stan|do el secondo in capo di tavola fermo, li altri 4 si tramuteranno et in 24 modi et stando fermo el tertio, li altri 4 si mutano pure in 24 modi, et stando pure el quarto in capo, li altri 4 medesimamente si mutano in 24 modi, et stando poi el quinto in capo, li altri si mutano in 24 in muodo che in tucti si mutano in 120 modi. Per sapere li 6 homini in quanti modi si tramutano, multiplicha 120 modi per 6 homini, fa 720 et tanti sono li modi che li 6 homini si possono mutare. Et per sapere li 7 in quanti muodi, multiplicha 720 modi via 7 homini faranno 5040 per li modi che li 7 homini et per li modi delli 8, multiplicha 8 via 5040, fa 40320 per li modi delli 8 homini. Et per li modi delli 9 homini multiplicharai 9 via 40320, fa 362880 et in tanti modi, dirai, che si aconcino 9 homini. Per li 10 homini multiplicharai 10 via 362880, fa 3628800 et in tanti modi, dirai, che si raconcino li 10 homini. Hora per vedere quanto montavano questi pasti a uno grosso l'uno, partirai in 143r questo el numero delli modi perché, comme è dicto per omni modo, lui li | haveva a fare uno pasto, dunque li haveva a fare 3628800 pasti. Dunque parti-

1 possono] F possino 12–13 tramutano] F tramutono 27 tucti] F tucto 29 6] K septe | li 7] F li 7 homini

rai questo numero per 20 perché 20 grossi vagliono uno ducato. Dove partito questo numero dicto per 20, ne viene 181440 ducati et tanto montano li pasti, li quali haveva a fare questo hoste et haveva a farli loro per 40 ducati. Et così stati costoro al quanto tanto che havevono consumati più di 100 ducati di robba. El povero hoste, chiamato Picho Maschalzone, del suo poccho iuditio advedutosi, alla sanctità del papa ricorso. Li nostri fiorentini, mossi a conpassione che ad observare loro tali pacti non li bastava se havesse havuto 10 volte più robba che non haveva, da quello contracto lo liberorno et anchora li costò 100 ducati a farsi liberare. Siché qui si mostra quanto giovi et nuoca alle volte lo essere et non essere experto nella scientia mathematicha, che se questo hoste fussi stato intelligente, non sarebbe stato così preso al bocchone di 40 ducati. Siché li hosti tenghino benissimo questo bello caso a memoria et non saranno mai ingannati.

Caso XVIIo di uno monasterio di monache 143v

Fu già nelle parte di ponente nella insola di Anglia uno monasterio di monache el quale, havendo pocha custodia sopra di sé, et per questo ne seguiva una largissima libertà di fare tucto quello che a lloro piaceva. Onde queste poverelle da qualcheuno, comme interviene, instighate, vedendosi in tale libertà, a pocho a pocho cominciorno a fare captivo habito et di uno in altro seguendo, vennono in tanta infamia apresso el vulgo che non altrimenti era tenuto questo monasterio che il publico luogho. Onde venendo tale abominevole caso alli orecchi della sacrata corona del re, deliberò sua maiestà tale inconveniente torre via, perché cognosciva tal caso essere di molta infamia al suo paese et inoltra alli sua popoli dava cativissimo nome. Et volendo a tale nefandissimo scelere rimediare, fece ad sé venire lo arcivescovo di quella ciptà, dicendoli che intendeva si rimediasse a tanto inhonestissimo acto et horrendo caso et che molto si maravigliava della persona Reverenda sua | che prima non vi havessi per sé medesima preso 144r
qualche buono ordine, protestandoli che se infra certo termine non vi havesse dato rimedio, faria tale monesterio desolare et le monache tucte metterebbe dove di stare meritavono. Et mostrato allo arcivescovo che questa cosa molto li era a core, a che il prefato arcivescovo tucto turbato et perterrito, allegate a sua satisfactione quelle scuse che conveniente li parevono, permise al tucto alla maiestà del sacrato re porre a ciò oportuno rimedio et debito fine et da quella preso debita licentia, prestamente al dicto monesterio si transferì et ad sé facta venire la badessa, aspramente la riprese, mostrandoli in quanto captivo con-

2–3 li quali] F che 5 Picho Maschalzone] F Picpo Mascalzoni 9 nuoca] K nocua 11 sarebbe] F saria 15 Anglia] F Aglia | monasterio] F monistero 20 tanta] F tale 29 rimedio] F octavo

cepto era apresso el preposto. Et della sacra corona recitandoli le parolle che dal re li erano sute exposte, a che la badessa non hebbe ardire niente rispondere comme quella che in verità cognosceva havere fortemente errato. Seguitò più avanti lo arcivescovo dicendo che per volere in questa sacra religione potere
144v habitare li bisognava mutarsi di vita | et di costumi in modo et forma che la già divulgata abominatione con sanctità, digiuni et vigilie ricuperassi. Il che quantunche li paressi duro, per fugire il periculo eminente che si vedeva, tucte permisono di farlo. Seguitò l'archiepiscopo: "volendo ciò fare bisogna tegnate quelli modi et via che io vi dirò. È necessario facciate uno novo habitaculo qual sia più strecto et mancho periculoso a fare quello che per insino adesso havvete facto. La ragione è che se voi state in questo monesterio, la fama captiva quale voi havvete mai si spegnerà perché quando bene non facciate niente, sarete ad omni modo inputate et quelli quali vi hanno havuta praticha, si potranno male distorre. Et per levare via tucte le cause io vi farò murare uno bello habitaculo et quivi meglio poterete vostra perduta fama ricuperare". Le quali cose, benché malvolentieri lo facessino pure, la vergogna et il timore le fece tucte consentire. Ordinò adunque el reverendissimo archiepiscopo uno bello et forte monasterio
145r nella for|ma che qui da parte lo vedi disegnato et è quadro et per omni faccia sono le camere del dormentorio di loro habitatione et perché le monache erano per numero 24 con la badessa, fece fare 8 lecta non già di piuma ma sacchoni in terra. Et in questo monesterio fece el dormentorio atorno atorno comme vedi, senza tramezi alcuni et insu 2

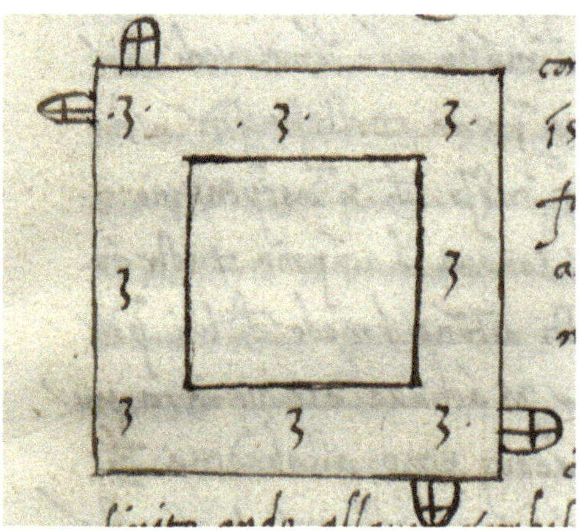

FIGURE 7
Drawing on folio 145r

canti, comme vedi. Fece insu omni canto dua finestrelle piccole, ferrate in modo che da omni canto facilmente poteva vedere 2 faccie et comme l'hebbe

5 bisognava] F bisogna 7 tucte] F *omissis* 21 atorno atorno] F intorno intorno

finito, andò alla venerabile badessa et disseli comme haveva edificato uno bello
monesterio più adorno et spatioso che il primo. Onde li acese l'animo a llei et
a tucte l'altre di andarvi dentro ad habitare. Et uno giorno diputato tucte le
rinchiuse in quella gabbia che comme vedi non haveva se non il dormento-
rio senza alcuno | tramezo né luogho dove nascondere si potessino. Comme 145v
l'hebbe là dentro, fece murare subito la porta in modo che li ucelli apena
entrare et uscire vi potevano, di che molto mal contente furno et di tale habita-
tione, parendoli havere mal facto, ma meglio potettero, a patientia si aseptorno.
Comandò loro il prefato arcivescovo che ciaschuna si aseptassi in questo modo
comme vedi, che quando lui guardassi per una delle finestre, vedessi essere 9
per omni verso che si aseptassino 3 per lecto et che comme sonava la cam-
pana, tucte fussino alli loro lecti o intorno, in modo che lui potessi vedere se
ve ne manchava o se ve n'era più che l'ordinario, ponendovi pena alla badessa
che se non faceva tale legge observare, la farebbe tucti li sua giorni vivere a
pane et aqua. Et la badessa da simile commento perterrita, alle sue sorelle
tucto riferì, pregandole volessino essere contente observare questo precepto.
Et tucte li promissono farlo. Et così spesse volte el giorno lo arcivescovo man-
dava a visitare queste poverelle. Et erano 2 | che l'uno stava a uno capo et l'altro 146r
al'altro capo et per le finestre miravono se erano 9 per omni verso. Così stan-
dosi le poverelle monache, ducevano loro vita molto tribulata et afflicta, stando
in digiuni, penitentie et vigilie continuante. Li loro amici quali erano privati di
poterle vedere et parlare, altro pensavano mai se non modo di potere loro man-
dare qualche messo o inbasciata, et pensato sopra ciò, non trovorno altro modo
se non una nocte entrare in dicto monesterio. Et ordinata una scala di fune,
uno di loro che era guasto, bene bene comme si debbe si misse ad entrare nel
monesterio prefato di nocte tempo. Et entrato dentro, fu dalle sante monache
molto volentieri visto et maxime da quella che teneva del suo core le chiave.
Et tucte particularmente a costui si racomandavono, contandoli la misera vita
che lo arcivescovo li faceva tenere. Lo amico innamorato, con buone et con-
fortative parole da parte di tucti li loro amici le salutò, dicendoli che di niente
dubitassi|no et haveva seco portato di boni capponi et trebbiani. Et con una 146v
corda tucta quella nocte atesono a tirare su robba perché lo arcivescovo li dava
pocho altro che pane et aqua. Et poi apropinquandosi al dì, la badessa disse:
"bisogna che voi vi partiate perché non habbiamo luogho da potervi nascon-
dere. Et se lo arcivescovo ci trovassi più che 9 per omni verso, ci mandarebbe
tucte per mala via". Colui rispose et disse: "non dubitate che io voglio starmi
con voi 15 giorni et faremo in modo che quando verrà la rasegna, non saremo

8 a patientia si aseptorno] F si aseptorno a patientia | potettero] F poterno

più che 9 per omni verso, ma bisogna mi prestiate una tonnicha che io mi vesta comme monacha", il che presto li fu data. Si dimanda: venendo colui a rasignare, in che modo faranno con questo uno più a essere 9 per omni verso? Si dà questo modo: comme vedi qui in margine signata, benché vari modi si potrà trovare da te medesimo, ma saria troppo lungho. Solo ti dico che costui di mano in mano, hora l'uno et hora l'altro, vi condusse 11 delli sua compagni et lui uno a 12. Più non se ne può mectere

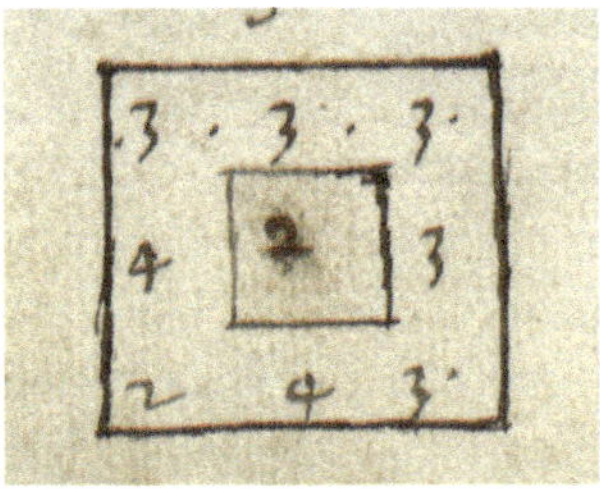

FIGURE 8
Drawing on folio 146v

147r che 12 a fare che a lo arcivescovo torni el conto che sempre sieno 9 per omni verso. Qui in figura si mosterà 12 finestre et comme si aconcia a crescere uno et 2, 3, 4, 5, 6, 7, 8, 9, 10, 11, 12. Et dentro nel quadro sarà signato el numero delli homini che vi crescie et vedrai li modi che se ne può molti più che questo trovare. Ma tale brigha voglio lassare al lectore che, immaginando con li quarteruoli, potrà molti muodi investigare, ma saria fastidioso tucti reciptare qui. Advenne che uno fu visto entrare in dicto monesterio et fu raportato allo arcivescovo et benché non lo credessi, per più cautela, col braccio del re, fece mettere uno bando che qualunche entrassi nel monesterio, o di dì o di nocte, o a quello si apressassi a 200 braccia, li mecteva pena la testa. Onde questi giovani che stimavono la vita dissono alle monache non volere più venirvi, ma che le monache con più facilità potevano fuora uscire et andare a trovarli.
147v Li resposono lo farebbono molto volentieri, ma che li dessono | el modo di potersi aseptare che el conto tornassi di 9 per omini verso quando ne fussi fora una o 2 o 3, etc. A che loro dissono che non ne poteva uscire più che 6 et lassorno scripto loro el modo comme si havevono aseptare quando ne era fuora 1, 2, 3, 4, 5, 6. Comme vedi qui in figura da canto che ti porrò 6 quadri, ma oltra a questo più assai delli modi si può trovare, ma questi basteranno alla solutione del caso. Et volendone più, farai comme ti ho dicto di sopra. Le figure delle donne che escano, haranno uno D di sopra, comme vedi, et dentro sarà il numero di quante escie fuora. Onde per abreviare diremo che

4–5 da te medesimo] F omissis 9 finestre] F figure 25 basteranno] F bastino

comme tu fai di 9 per omni verso, così puoi fare di 10, di 12, di 20, etc. Et così poi proporre la decima figura et dire: "cava 12 overo 15 di questa et fa che restino 9 per omni verso et molti belli casi". Hora perché fu dicto allo arcivescovo comme loro uscivano fora, di che lo fece intendere alla badessa che se a tale inconveniente non prevedeva che tucte sarebbono vituperate et presi et menate al postribulo | et che vi terrebbe le guardie onde al tucto fuorno 148r
privatele monache di omni comodità et piacere. Al che per disperate cominciorno dentro a fare el peggio che potevano et quando mandava a rasignare, si aconciavono in modo che erano quando 10, quando 11 et quando 12 per faccia et 8 et 7 et 6. Si domanda el modo, quale sarà qui in figura, per la qualcosa colui che rasegnava diceva essere più che l'ordinario et contendeva con epso. Loro dicevano: "chiamaci a nomi tucte et vedrai che non siamo più né mancho". Per questo fuorno in grandissima contentione et messer lo arcivescovo con la sua maliditione li lassò la briglia in sul collo che facessino quello volevano et acioché più copertamente lo facessino, cominciò lui con li sua prelati a usare con le sancte monache la buona charità et in quello modo levorno via tucte l'eresie et scandoli che prima vi erano. Hora li modi che loro possono crescere saranno qui in margine, e' quali sono 3 che non si possono mostrare più che 12 per faccia, sendo 24, et saranno signate: A, B, C. El modo che si tiene ad essere mancho di 9, cioè 8, | 7, 6 per faccia che mancho di 6 148v
non si può, non crescendo et diminuendo del numero delle 24. Sarà le 3 lettere dello alfabeto, cioè: E, F, G. Et questo basta quanto al presente caso che, benché sia prolisso, tamen è di piacere et bello, perhò non te increscha di leggierlo.

Caso xviiio di una croce che per omni verso è 9 et levarne et porne

Et da questa forza nasce uno altro caso et bello, el quale è questo: poni in tavola una croce simile a questa che è qui da parte vedi in figura et di grossi o quattrini, comme vuoi, el cui capo è A et el piè B et le braccia una C, l'altra D. Et dice in questa forma: "io comincio a contare da piè al capo et sono 9 et contando dal piè, il braccio C sono 9 et similmente contando dal piè et andando verso el braccio D, sono similmente 9. Hora io ne voglio levare 2 et faranno el medesimo effecto." Si dimanda in questo caso comme si fa. Dirai che si può levare, nonché 2, 4 et fassi così: | volendone levare 2, levane da omni braccio una che 149r
vedi che sono 3 per braccio, dunque ti resterà dua per braccio. Et contando al modo dicto, non sarebbe più che 8. Et perché sieno 9, farai così: manderai il braccio più su uno puncto comme nella seconda figura. Vedi che nella prima

36 comme] F comme vedi

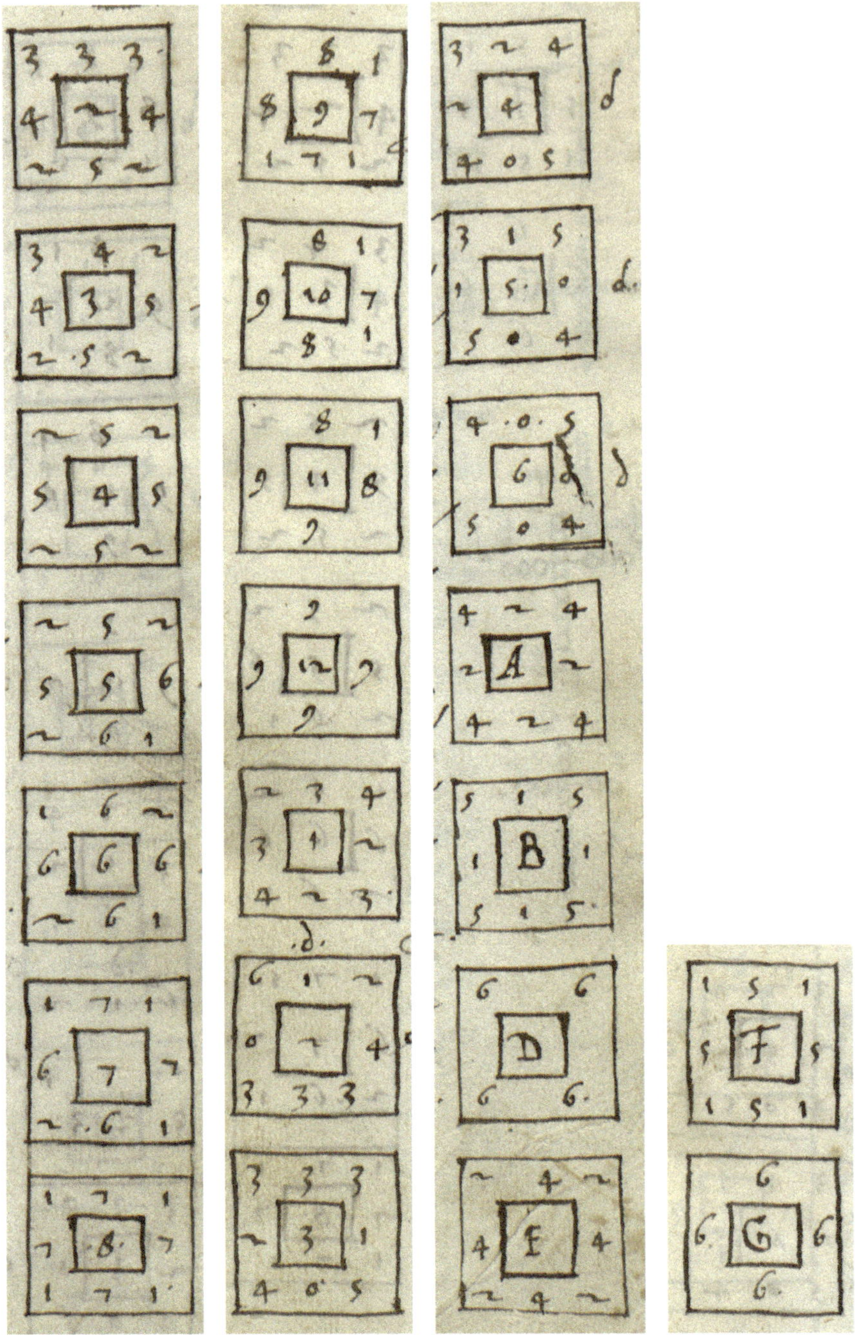

FIGURES 9–12 Drawings on folio 147r–148v

figura le braccia cominciano al sexto puncto, facendosi da piè et nella seconda figura le braccia cominciano dal septimo puncto et hai cavatone 2 grossi o quattrini o quarteruoli comme sono.

Et volendone cavare 2 altre, leverai pure dalle braccia et manderale uno puncto più su et sarà una moneta per braccio, comme vedi nella tertia figura qui da canto. Et più che 4 in simili figure non è possibile levarne, ma quando proponessi magior figura di 10 o 20 per omni verso, potrebbe levare più, secondo la grandeza della figura.

Et così similiter, comme si può levare, così si può agiugnere et comme in nel levare tu levi dalle braccia et mandile più su, così nel porre, ponile alle braccia et ritirale in giù, comme vedi nella quarta figura qui | dal lato dove è la croce, è 149v
4 per ciaschuno braccio et volendo porvene più, poni alle braccia et tirale uno punto più in giù et potrai molto più porre che levare che ne potrai porre fine alla somma di 8, cioè 4 per braccio, comme vedi nella quinta figura designata. Et molti varii muodi poterai proporre in questo modo.

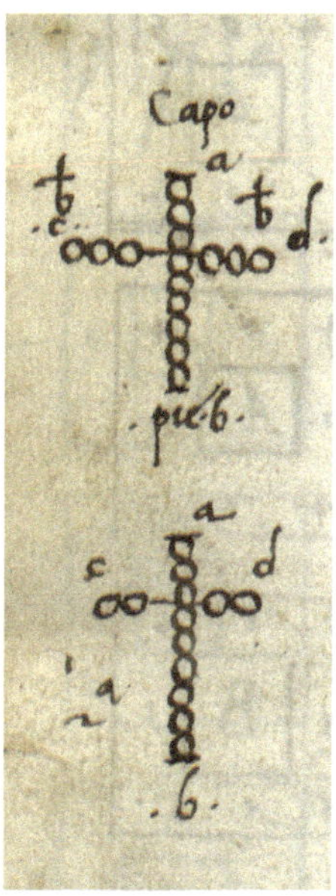

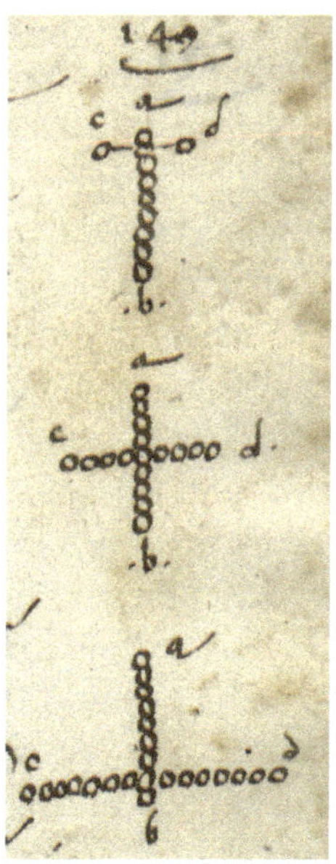

FIGURES 13–14 Drawings on folio 148v–149r

5 per braccio] F perché 12 volendo] K volendone

Caso XVIIIIo del modo di adoppiare uno granello di grano et durare uno mese

Sono molti che dicano: "io ti voglio dare 100 ducati se tu vuoi comminciare el primo dì a darmi uno granello di grano, el secondo 2, el tertio 4 et subcessive omni dì radoppiare el giorno antecedente. Et voglio che tu duri 15 giorni a fare questo". In questo caso dico chi dessi li 100 ducati perderebbe in grosso, comme manifestamente si può vedere, che adoppiando sino a dì 15, non fornirebbe in capo di dì 15, sommando tucto insiemi uno staro di grano. Et a ciò meglio tu possa fare questo conto, darò breve regola: a sommare et 15 et 150r 20 et 30, etc. farai così: piglia tucta la somma di 2 | o di 3 o di 4, comme vuoi, et secondo che tu hai adoppiare, a tucta la somma giugni uno et poi multiplicha tucto el numero in sé. Et farà la somma di altretanti di quanti tu hai preso la somma, comme se tu havessi preso la somma di 3. Di poi la multiplichatione ti darà la somma di 6 giorni. Hora diciamo che tu voglia adoppiare insino in 32 giorni, omni dì adoppierà l'antecedente giorno. Tu per fare presto piglia la somma di 8 giorni. Potresti, non ti bastando l'animo di 8 giorni, fare di 2 et poi di 4, ma questa fia più breve. Essendo tu buono abachista, ti fia facile. Piglerai adunque la somma di 8 giorni che sarà tucta 255 et a questo 255 poni uno, farà 256 et questo multiplicha in sé, farà 65536, del quale cava uno, resterà 65535 per la somma delli granelli di 16 giorni. Et volendo tu sapere la somma di 32, poni a 65535 1, farà 65536 et questo multiplica in sé medesimo, farà 4294967296, del quale cava uno, resterà 4294967295 per la somma delli granelli di 32 giorni. Et volendo tu sapere quanti stara sono alla nostra 150v misura fiorentina, | partirai questo numero per 345600 et, se bene noti, tanti granelli vanno in uno staio di grano che 'l sia la verità. Tu sai che in uno 1/8 di oncia vi va 72 grani, perché 24 grani sono 1 denaro peso et 3 danari pesi sono 1/8. Et dirai: 3 via 24 farà grani 72 per 1/8 et per sapere quanti grani vedi in una oncia, multiplicha 8 via 72, fa 576. Et poi, per sapere quanti granelli va in una libra, multiplicha 576 via 12, fa 6912 et tanti granelli di grano vanno in una libbra fiorentina. Hora tu sai che uno staio si fa comunemente libre 50. Et per sapere quanti granelli di grano va in uno staio, multiplicha 5912 via 50, troverrai che fa 345600 et tanti vanno in uno staro. Et volendo intendere quanti stara sono 4294967295 granelli, partili per 345600 et troverrai che sono staia 12427 et avanza 196095 granelli. Hora potrai valutare lo staio che uno tempo per l'altro si sole valutare soldi XX lo staio, siché tu potrai valutare 12427 staia, saranno tante libre et riducta a ducati d'oro troverrai che 151r sono 1775 ducati d'oro et libre 2, siché vedi chi ne | fa meglio et perhò in si-

7 sino] K, F sine 8 staro] K straro 18 255] F 225 19 256] F 226 29 vanno] F va

mile caso sarai advertito che tu non sia colto al bocchone. Et con questa regola sempre potrai calculare et 20 et 30 et 40 giorni secondo li casi che saranno proposti.

Caso xxo di uno altro modo del radoppiare uno granello di grano

Et nota che el radoppiare uno granello di grano si può fare in 2 modi. L'uno è el modo che sopra ho mostro et l'altro sarà questo che diremo, cioè che si radoppia sempre tucti li granelli che per lo passato hai havuto. Comme tu dirai: "dammi el primo dì uno granello", el secondo 2, el tertio radoppia tucti li precedenti, cioè uno et 2 fanno 3, doppiato fa 6, et il quarto dì te ne darà 18, etc. Hora questo modo è magiore qualità assai che non è la superiore. Et anchora a questo fare ti darò breve regola: diciamo che, comme nella superiore, tu voglia radoppiare 32 dì, adoppiando 32 giorni, adoppiando omni dì quello che tu havessi havuto nelli dì preteriti, a te bisogna questo modo pigliare la somma di | quanti dì tu vuoi, inclusive di tucti et multiplicha dicta somma [151v] in sé. Ti gitterà la somma di altretanti dì, mancho uno che quello che pigliasti. Comme se tu pigli la somma di 6 giorni et quella multiplichassi in sé, ti getterà la somma di quelli 6 et di altretanti mancho uno, cioè di 5 altri che in tucto harai la somma di 11 giorni. Et poi se questa somma intera multiplichassi in sé, ti darebbe somma di altretanti dì mancho uno che questo sono 11, dunque ti darà la somma di 21 dì in tucto. Adesso diciamo che vogliamo adoppiare 32 dì, comme nella superiore. Piglerai, comme nella precedente, la somma di 8 giorni che, se bene noti, te ne dà uno, el secondo 2, el tertio 6, el quarto 18, el quinto 54, el sexto 162, el septimo 486 et lo ottavo 1458, che se tucto sommerai insiemi, farà 2187. Hora multiplicharai questa somma insiemi in sé medesima, farà 4782969 et questo sarà la somma di 15 giorni. Poi multiplicha questa somma delli 15 giorni in sé medesima, farà 22876792454961 che sono 14 figure et questa è la somma di 15 giorni et di 14 se|guenti che tucti sono 29. [152r] Et noi vorremmo sapere la somma di 32. Hora questo resto si può fare in più modi, ma questo mi pare più expedito. Cioè radoppia la somma delli 29 giorni dicta di sopra, farà 45753584909922, et questo acozato con 22876792454961 farà 68630377364883 et questa sarà la somma di 30 giorni apunto. Hora radoppi questa ultima somma et poi acoza tucto insieme, cioè la somma delli 30 giorni col suo doppio, farà 205891132094649 et tucta questa è la somma delli 31 giorni. Poi adoppia questa somma et agiugni insiemi el dicto doppio con la somma delli 31 dì, farà 617673396283947. Et questa sarà la somma di 32 giorni, apunto.

---

24 2187] K, F 2182 31 30 giorni apunto] F 30 giorni 33 205891132094649] K, F 207891132094 649

Potresti fare questo medesimo per altro modo, comme adesso dirò. A comminciare a sommare 5 giorni inclusive quanti granelli tu faresti che il primo giorno è uno, el secondo sono 2, el tertio 6, el quarto 18, el quinto 54, che tucto insiemi acozato fa 81. Et questo 81 è la somma di 5 giorni. Hora multiplicha 81 in sé,
152v fa 6561 et questa è | la somma di 9 giorni, comme è dimostrato. Et se tu multiplicharai 6561 in sé, farà 43046721 et questo è la somma di 9 giorni et di 8 seguenti che in tucto harai la somma di 17 giorni. Hora, se tu multiplicharai la somma di 17 giorni in sé medesime, farà 1853020188851840 et questa è la somma di 17 giorni dicti et di 16 che seguitano, che in tucto farà somma di 33 giorni. Hora tu vorresti doppiare 32 giorni et habbiamo uno più che el bisogno. Et per venire alla notitia delli 32, hai a partire tucta la somma delli 32 giorni per 3 perché in quella somma vi si contiene 3 volte el doppio di 32 giorni, perché comme tu se' a 32 dì et voglia per la regola sopra dicta vedere quello va in 33 giorni, tu adoppierai el numero delli 32 et sopra el doppio giugnerai el numero delli 32 giorni. Eccho che in 33 dì vi si contiene 3 volte el numero delli 32, dove se tu partirai el numero sopra dicto per 3, ne verrà 617673396283947 che comme vedi sono 15 figure et vedi che si riscontra con la regola di sopra.
153r Et piglierai | di questi 2 modi quello ti è più facile et ancora secondo li casi che ti saranno preposti et vedi manifesto che questo caso è magiore che lo antecedente che in quello la somma delli 32 giorni era 4294967295 che in tucto sono 10 figure et queste, comme vedi, sono 15. Hora, se tu vorrai vedere quanti stara di grano sono questi, partirai la somma dicta per 345600 et harai quanti staia sono et nota che quando tu venissi a simili partiti che tu patteggi chiaro a che modo tu hai radoppiare o in questo modo o nel superiore che, comme vedi, ci è grandissima differentia. Sono questi simili casi di grande importantia et paiono di piccola, comme nel caso XVI superiore delli 10 homini in quanti modi a pasto si possono aseptare, che uno babbione saria colto al bocchone et rimarebbe alla trappola comme fa el topo, siché la mathematicha scientia è bona haverla in omni conto che spesso è suto danno a qualcheuno per non essere buono mathematicho, comme advenne a quello povero hoste.

153v **Caso XXIO di uno che adoppiassi una quantità**

Puossi proporre uno bello et stupendo caso fra la brighata et è di grandissimo piacere a chi ne ha notitia, quale è questo. Dirai a uno che pigli uno monte di quarteruoli o di quattrini a te incogniti et che cominci da che numero lui

8 giorni] F case | 1853020188851840] K, F 1853020188851841 11 notitia delli 32] F notitia
18 secondo] K secondo, etc. 27 aseptare] F asectare

vole o da 1, o 2, o 3, a suo libito, solo ti dicha da che numero lui comincia et che poi pigli el doppio del primo et vadi continuando, doppiando sempre lo antecedente et metta in monte di mano in mano tucto. Et non vuoi comme nella superiore che ti dica el numero delle poste, ma solo ti dica l'ultima posta che lui mecte et tu in uno instante ti obligherai a sapere dire quanto sia el monte tucto che veramente fia tenuto mirabile. A questo fare ti voglio dare regola generale. Quando lui cominciassi da 100 o da 1000, a suo modo, et durassi tucto uno giorno a doppiare, presto et con velocità lo farai. Tieni questa regola: cava el primo termine del'ultimo, cioè cava el numero donde lui cominciò a doppiare de|l'ultimo che lui finisce. Et quello che rimane, agiugni sopra l'ultimo termine 154r
che lui ti dirà et tanto fia la dicta somma et mai ti fallirà. Comme se tu cominciassi da uno et poi 2, 4, 8, 16, 32, 64, 128, 256, 512, 1024, 2048, etc. Dico che in questo caso tu cavi el primo numero che è primo del'ultimo che è 2048, resterà 2047, dove acoza insiemi questo 2047 con l'ultimo numero, cioè con 2048, farà 4095, per tucta la somma inclusive da uno insino in 2048, doppiando solamente. Et se tu havessi cominciato da 3 et dicessi 3 et poi 6, 12, 24, 48, 96, 192, 384, 768, 1536, etc. In questo ancora dico che tu cavi el primo termine che è 3 et del'ultimo numero, cioè di 1536, resterà 1533 et questo 1533 agiugni sopra l'ultimo termine, cioè sopra 1536 farà 3069 et questa è la somma di tucti li doppiamenti da 3 sino a 1536 et così se lui pigliassi numero rotto o intero o rotto in qualunche modo, piglierà sempre tenendo la data regola, ti riuscirà. Et nota

[calcoli a margine che accompagnano la demostratio]*

che con questa regola et ordine tu potrai fare cose stupende che se fusse uno 154v
monte di grano alto 100 braccio, tu ti poi obligare a sapere dire quanti staia è decto grano et quanti granelli saranno. Et con questa regola si potrebbe numerare tucte le arene del mare quanti granelli sono et molte cose stupende che tucte derivano dalla forza et virtù del numero et sua proportione. Et essendo intrato in questo, voglio dare qualche altra regola bella di simili casi comme questa.

---

15 per tucta la somma inclusive da uno insino in 2048] F *omissis* 18 cioè] K ciò 20 sino] F insino

---

* I calcoli mancano in F

Caso XXIIo quando uno andassi rinterzando

Et se per caso uno dicessi volere cominciare da uno numero a suo libito et poi, di mano in mano, reinterzare insino a quanto a llui pare, a questo anchora ti darò una regola da potere tucto con presteza racorre. Sempre in questo caso cava el primo numero che lui prese del'ultimo et quello che resta dividilo per 2 et quello che ne viene, poni sopra l'ultimo numero et quella sarà la somma di tucti li decti numeri.

1
3
9
27
81
243
729
2187
2187 ultimo
1 primo
2186 parti per 2
viene 1093
2187
3280

155r Comme se uno havessi cominciato da uno et dicessi uno et poi 3, 9, 27, 81, 243, 729, 2187, etc. Dico che cavi uno primo numero del'ultimo che è 2187, resterà 2186 del quale piglierai la metà, che è 1023, et questo poni sopra l'ultimo termine che è 2187, farà 3280 apuncto, per la somma di tucti li numeri da uno sino a 2187 per numero tertiario, ascendendo comme puoi per te medesimo provare. Et se lui si fussi cominciato da 20 et dicto prima 20 et poi 60, 180, 540, 1620, 4860, dico che cavi 20 primo numero di 4860, ultimo resterà 4840 et di questo piglierai la metà che fia 2420. Hora dico che agiungha questo 2420 al'ultimo termine, cioè a 4860, farà 7280 per tucti li numeri da 20 insino 4860 inclusive, comme è dicto. Et se tu cominciassi dal rotto solo o dal rotto et intero insiemi, tenendo la data regola, sempre ti riuscirà benissimo, quando durassi tucto uno giorno a contare, solamente dirai: vuoi sapere el numero donde lui si comincia el numero dove lui finisce et non

---

14 di] K *omissis*

20
60
180
540
1620
4860
4860 ultimo
20 primo
4840 parti per 2
viene 2420
4860
7280

altro, comme tu da te medesimo potrai experimentare. 155v

Caso XXIIIo del modo di uno che per numero quaterno ascendessi

Et se, per mutare modo, tu volessi racorre con presteza uno numero di uno che andassi per numero quaternario, comme uno, 4, 16, etc., anchora a questo ti darò regola generale in quanti modi uno volessi ascendere che havendo a mente questa gentile regoletta innanzi a uno signore overo gran maestro, ti faranno sempre et utile et honore che sono di pocha faticha el tenerle a mente, che una regola che qui inmediate ti dirò ti servirà in tucti e' modi che uno volessi procedere. Hora farai così: cava el primo termine di questo ultimo et quello resta parti per 1 manco che 'l numero per il quale l'amico tuo ascende comme in questa: lui ascende per 4 et tu parti per 3 quello ti resta, cavando lo primo termine dello ultimo et quello che ne viene, agiugnilo sopra l'ultimo numero o termine, et quella sarà tucta la somma, o rotto o intero, o rotti o interi, insi|emi 156r
o in omni et qualunche modo che tu vuoi.

1
4
16
64
256
1024

6 regoletta] F regola 8 dirò] F darò 11 cavando] F l'avanço

1024 ultimo
1 primo
1023 parti per 3
viene 341
1365

Comme se lui comminciassi da uno et dicessi: uno et poi 4, 16, 64, 256, 1024, etc., dico che cavi uno primo termine di 1024, ultimo termine resterà 1023 et di questo piglia il tertio, che sarà 341, et questo poni sopra 1024, farà 1365, per tucta la somma di decti numeri quadruplati et comincise donde vole. Et se l'amico volessi andare per numero di 5, cioè quinquplare sempre el suo numero, comme dire 15, 25, 125, 625, 3125, etc., caverai uno primo termine del'ultimo, cioè di 3125, resterà 3124 et di questo, perché l'amico è andato quinquplando, piglia el quarto, sarà 781 et questa quarta parte, cioè 781, agiugnerai sopra l'ultimo termine, cioè 3045, farà 3906 per tucta la somma delli inclusi numeri. Et se lui andassi per 6 et tu piglieresti el quinto, et se per 7 et tu el sexto et se per 8 et tu el septimo, di mano in mano piglierai uno mancho che il numero piglia lui. Se bene lui andassi per 100 o per 1000 et tu pigleresti uno mancho di lui, et
156v sempre ti | riuscirà benissimo et con presteza comme tu da te medesimo puoi experimentando provare.

Caso 24 di uno altro modo di racorre

Et quando ti fussi dicto: "io ho facto uno monte di monete, sia di che quantità si voglia, o sassi o quarteruoli o quomodo cumque sit" et dicessi: "io ho messo tucti li numeri pari che sono da 2 sino a 20 o a 24, etc.". In questo caso et in simili terrai questa regola: piglia la metà del'ultimo termine et quella metà multiplicha per uno più et farà dicta somma. Comme se uno havessi messo in monte 2, 4, 6, 8, 10, 12, 14, 16, 18, 20, in questa piglia la 1/2 di 20 ultimo che è 10 et questo multiplicha via uno più, cioè via 11, farà 110. Et questa è la dicta somma, inclusive di tucti li prenominati numeri.

Et se lui dicessi havesse messi tucti li numeri caffo overo dispari che sono da uno fino a 21, comme 1, 3, 5, 7, 9, 11, 13, 15, 17, 19, 21, terrai questa altra regola: fa 2 parte del'ultimo numero maggiore che si può senza rompere le unità et per la magiore [calcoli a margine]

2 resterà] F *omissis* 7 perché] F che 11 piglierai] F piglia 12 uno] F *omissis* 22 via] F per 24–25 da uno fino a 21, comme] F *omissis*

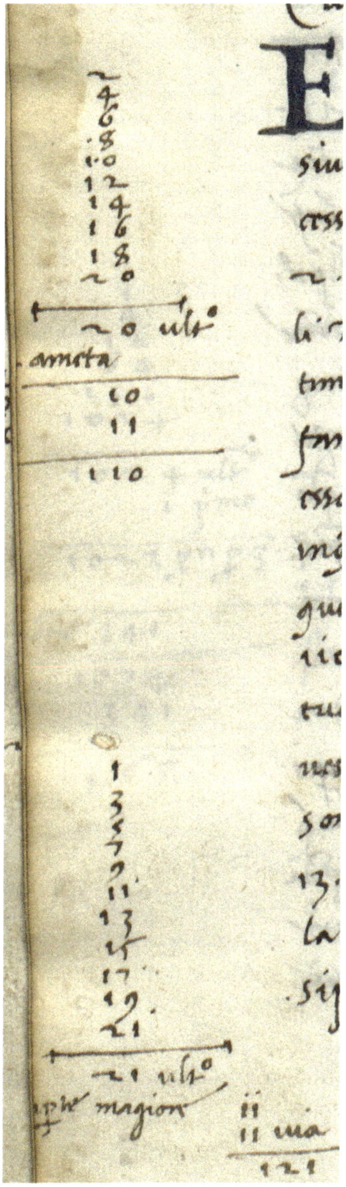

FIGURE 15
Marginal notes on folio 156^{v}

parte multiplicha in sé et quello fa, sarà dicta somma. Comme in questo ultimo termine è 21, factone 2 parte, le maggiore sirà una 10, l'altra 11. Presa adunque la maggiore et multiplichata in sé, farà 121 et questa è la somma delli prenominati numeri. Hai dunque la regola quando saranno pari et quando dispari. Ora nota una regola a tucti generale, o pari o dispari, non ti darà noia. Con questo

seguente modo tucte le potrai absolvere. Farai così: piglia el numero delli termini quanti sono et di quelli prendi la metà et con questa metà multiplicha el primo termine posto sopra l'ultimo et questa sarà sempre dicta somma. Comme nella superiore di 2, 4, 6, 8, 10, 12, 14, 16, 18, 20, tu vedi che tucti li numeri sono 10. Dunque piglia 5 et poi poni el primo termine che è 2 sopra l'ultimo quale è 20, farà 22, quale multiplichato per 5 farà 110 per tucta dicta somma. Et nello altro caso di 1, 3, 5, 7, 9, 11, 13, 15, 17, 19, 21 tu vedi che li numeri sono 11 del
157v quale la metà è 5 1/2. Dunque poni uno | primo termine sopra l'ultimo, cioè 21, farà 22, et questo multiplicha via 5 1/2 fa 121 per la somma di tucti li prenominati numeri. Et mai ti fallirà questa regola et comincisi l'amico da che numero vuole, purché non vi sia rotto che quando vi fussi rotto non potresti per questa regola ritrovarlo. Questa serve per li numeri pari et dispari, comme vedi.

Caso 25 di uno altro modo di racorre

Se l'amico dicessi: "io mi comincio dalla unità et ascendo arismetice et continue", cioè dico: 1, 2, 3, 4, 5, 6, 7, 8, 9, 10, 11, 12, etc., anchora a questo con presteza ti darò regola. Et comme in dua modi può terminare, comme in pari et caffo, così in 2 modi si può absolvere con presteza et prima. Quando terminassi in numero pari, piglia la metà del'ultimo numero et quella multiplicha via uno più del'ultimo termine et questa sarà tucta dicta somma comme in questo: 1, 2, 3, 4, 5, 6, 7, 8, 9, 10, 11, 12. Piglia la metà 12 ultimo è 6, questa multiplicha per
158r 1 più di 12, cioè per 13, farà 78, per la somma di tucti li preno|minati numeri quando sieno pari. Ma quando sieno dispari ti servirà questa altra regola: fa sempre dello ultimo termine 2 parti maggiori che tu puoi, non rompendo la unità. Et poi con la magiore et tu multiplicha dicto ultimo termine et quello sarà apuncto la somma. Et mai falla comme a dire 1, 2, 3, 4, 5, 6, 7, 8, 9, 10, 11, 12, 13, fa di 13 2 parti, le magiori sirà l'una 7, l'altra 6. Piglia adunque la magiore che è 7 et con questa multiplicha lo ultimo numero quale è 13, farà 7 via 13, 91. Et questa è la somma di tucti li prenominati numeri inclusive, cominciando dalla unità et procedendo comme è dicto. Et anchora per una sola regola potrai sommare tucti li numeri et termini in pari o caffo, non porta, sapendo li 2 extremi. Farai così: piglia sempre la metà dello ultimo numero et a dicto numero agiugni uno et multiplicha per dicta metà et farai dicta somma. Comme nella superiore, dove 13 è ultimo numero, partilo per 1/2, è 6 et 1/2, et a dicto 13 poni uno,
158v farà 14 et questo 14 multiplicha per 6 1/2 | fa 91. Et così ti riuscirà in tucti i modi che la terminassi quando lui cominci dalla unità et proceda con la unità, sempre comme è dicto, ma quando cominciassi da altro numero che dalla unità,

6 quale] F che 14 mi comincio F] incomincio 18 numero] F termine

queste regole non ti serviranno, comme puoi provare. Et perhò ti voglio dare nel capitolo che seguirà una generale regola in tucti li modi che uno procedere volessi comme vederai et perhò nota.

Caso 260 a racorre in tucti i modi regola generale

In questo capitolo voglio dare regola generale del racorre tucti li numeri che ti fussino proposti. Et ascenda arismetice o per qual numero vuole o per 3 o per 4 o 5, ect., o pari o caffo che sia, o sieno rotti soli o interi con rotti o interi soli, la regola fia generale in omni et qualunche modo tu voglia procedere, non fa caso. Et tu per racorla presto et con velocità terrai questa seguente regola: poni sempre el primo termine overo numero in sul'ultimo et questa somma multiplicha per la metà del numero delle poste, cioè delli numeri che in tucto vi si contenghano. Et quello fa, sarà la somma di tucti li decti numeri inclusive. | Comme 159r
se tu volessi racorre questi numeri 1, 2, 3, 4, 5, 6, 7, 8, 9, 10, ect. Et questo modo è modo ordinario. Et saputo che lui procede ordinariamente arismetice, tu saprai che li termini tucti sono 10. Dunque poni uno termine primo sopra 10 ultimo termine, fa 11 et questo multiplicha per la metà delli termini che sono 10 la cui metà è 5. Dunque multiplicha 11 per 5, farà 55 et questa è tucta la dicta somma. Se tu volessi racorre questi numeri: 3, 5, 7, 9, 11, 13, 15, 17, 19, 21, la quale progressione arismetrice per 2, porrai el primo numero, che è 3 in sul'ultimo che è 21, fa 24 et questo 24 multiplicha per la 1/2 delli numeri o termini che sono 10. Dunque multiplicha 24 per 5, fa 120 per tucta la prenominata somma et nota che a te basta sapere el primo et l'ultimo termine solamente. Et per ritrovare con facilità li termini quanti sono, fa così: cava el primo termine del'ultimo et quello che rimane parti per numero della progressione et quello che ne viene sarà uno mancho del numero delli numeri comme nella superiore. El primo termine è 3 quale cav|ato di 21 resta 18 et questo parti per el numero della pro- 159v
gressione che in questo è 2. Dunque partito 18 per 2, ne viene 9 et questo 9 è uno mancho delli termini. Dunque agiugni uno a 9, fa 10 et questo è el numero delli sopradecti termini, comme vedi. Et così, se tu andassi per numero trino et cominciassi da 5 et dicessi 5, 8, 11, 14, 17, 20, 23, 26, 29, etc., prima per trovare el numero delli termini cava 5 di 29, resta 24, questo parti per el numero della progressione che è 3, ne viene 8, a questo poni uno, fa 9 et questo è el numero delli termini dal quale la metà è 4 1/2 et serba. Hora poni el primo termine sopra l'ultimo, fa 29 et 5, 34, et questo 34 multiplicha via 4 et 1/2, fa 153 et questo sarà tucta la prenominata somma. Et mai ti fallirà et vadi per che progressione vole.

2–3 uno procedere volessi] F uno volessi procedere 6 arismetice] F omissis 9 et con velocità terrai questa seguente regola] F *omissis* 15 per] F *omissis* 16 10] F 5 17 farà] F fa 20 1/2] F metà 22 a te] F *omissis* 24 parti] F *omissis*

Salvo che questa regola non serve alle multiplichationi di sopra dicte comme
a multiplichare per 3, per 4, et 5, etc., ma alle progressioni quomodo cumque
160r sit, ti farà honore et con questi preallegati casi potrai molti bellissimi | et vari
giuochi fundare et dare piacere alle brighate con mille nove inventioni et casi
formati per te medesimo.

Caso 27 d'uno altro modo di racorre

Et se uno dicessi: "io piglio, verbi gratia, 4 et poi continuo sempre al numero che io ho agiuncto la sua metà et mecto in monte comme prima, dico 4, poi 6, 9, 13 1/2", in questo caso dirai così: "sempre multiplicha el primo termine per 2 et questa somma leverai della multiplichatione del'ultimo termine in 3 et quello resta, sarà la somma di tucti li decti numeri". Dunque in questa multiplicha 4 primo termine per 2, fa 8, et poi multiplicha 13 1/2 via 3, fa 40 et 1/2. Hora cava di 40 et 1/2, 8, resta 32 1/2 et questa è la somma di tucti li decti numerii. Et se lui ponessi il tertio, sempre comme dire 9, 12, 16, etc., et tu multiplicha el primo per 3 et cavalo del quadruplato dell'ultimo et quella sarà la somma. Comme in questa: multiplicha 16 per 4, fa 64 et di questo cava 3 via 9, fa 27. Dunque
160v cavato 27 di 64, resta 37 et mai ti | fallirà. Et se lui ponessi el 1/4 et tu multiplicaresti el primo termine per 4 et cavarestillo della multiplichatione del'ultimo in 5 et haresti tucta la somma. Perhò nota che sempre tu harai a multiplichare el primo termine per il numero della sua ascensione in questo caso et caveralo della multiplicatione del'ultimo termine in uno più ch'el primo et sempre ti riuscirà. Se tucto uno giorno durassi a contare per che numero si voglia, mai ti fallirà. Hora tu hai tuccti li modi che si può procedere a numerare, a tucti hai la sua regola. Et quando tu harai questa regola alla memoria, ti farà in omni conto honore mirabilissimo.

6 27 F] 26 10 della] F dalla 13 32 1/2] F 32 21 della] F dalla 25 mirabilissimo] F mirabilissimo in omni luogho

Seguita la quarta et ultima parte del libro dicto: Giuochi Mathematici

Io veggho con l'occhio speculativo della mente nell'aria una fulgentissima nube, intorno intorno da lucidissime stelle circumdata. Et era di tanto fulgore che l'occhio mio corporale in ep|sa non haveva posanza. Et in quella nube dentro 161r vi era una speciosissima giovane di età matura, ma le sua innumere et mirabile belleza non potrei con penna explicare. Et lo aspecto suo mostrava quella havere in sé grandissimo inperio et dominio. Et era tucta vestita di candidissime veste et el capo suo di una aurea corona era hornato. Et stavasi a sedere in una bellissima sedia di bellissimi brocchati tucta circumdata, la qual sedia da quattro fulvi leoni sopra loro robustissimi humeri substenuta et portata era. Et la prelibata veneranda donna, nella sua dextra in mano uno aurato baculo teneva et nella sinistra una lucidissima spada. Et factasi al quanto alli miei occhi più vicina, viddila essere nel suo aspecto molto turbata, stando con la mente et animo suspesa. Et haveva intorno a sé molti homini nelli loro aspecti gravi et di somma scientia, col governo et consiglio de quali, secondo el mio juditio, si governava questa famosa regina. Donde io, comme quello che desiderava intendere quella | donna chi fussi et di che havesse dominio et quello che più desiavo 161v era intendere quale fusse la causa che sì fantastica et malenconica l'asistessi, me acostai in modo, alla prefata nube, che il parlare a uno delli astanti mi fu facile. Et apropinquatomi ad uno di quelli che nella predicta corte mi pareva ministro et nello aspecto humano et gentile, li dissi se la mia dimanda è licita, io harei caro sapere qual sia questa magnanima regina, il che lui mi manifestò. Et procedendo più oltra, li domandai se sapeva la causa che sì afflictta et mesta stare la faceva. A che mi rispose che tal domanda per anchora non era licita, ma se lo stare alquanto quivi non mi era molesto, che per me medesimo lo intenderia. Per lla qual cosa, essendo di vedere cose nuove, vago deli beni di quel locho non partirmi, che il fine di ciò vederia. Et stando quivi per alquanto spatio furono li occhi mia da magior splendore occupati, di modo che quasi me privò di potere più niente vedere, in modo che stetti alquanto con li occhi chiusi. Et | di poi, apertoli a pocho a pocho, la perduta vista ricuperai et 162r alquanto asicuratomi di potere alzare li occhi in quella parte dove era quella nova luce, li girai et viddi quello splendore procedere da una parvoletta nube che dallo altissimo celo, secondo el mio iuditio, procedeva. Perché io, levato lo animo dalla prima nube, solamente a mirare quella parvoletta et sua effecti ero intento. Et mirando quella, viddi che a pocho a pocho, suavemente inverso

32 asicuratomi] F riasicuratomi 36 suavemente] K suamente

di me veniva. Et era epsa parvoletta nube a guisa di una mandorla et da due grandissime aquile portata, una nella sua sommità et una a pié di dicta nube, et in quella si vedeva uno nobilissimo signore che di così mostrava suo regale aspecto. Et la sedia sopra la quale sedeva era di fulgentissimo auro conposta et fabbrichata et le sue veste erano di finissimi drappi et al suo sinistro lato una fulgente spada cinta teneva et nella dextra mano uno ramo di annosa et
162v viridissima quercia haveva et nella sinistra una palla tucta di puro oro. Et | alli sui piedi sedevano septe spetiose donne, tucte di bianchissimi vestimenti hornate. Et intorno a sé haveva molti homini che tucti nello aspecto mostravano essere nobili et di grandissima balia et potere. Et mentre stavo a contemplare la prefata nube, in uno instate viddila a quella prima nube magiore factasi vicina, con quella congiugnersi in modo diventorno una medesima. Per la qual cosa quella sapientissima regina, tolto da sé omni mestitia et rivestito el suo volto di inmensa letitia, levatosi ritta della sua sedia, prese quello magnifico homo per mano et al suo dextro lato volle che con la sua sedia se dimorassi a llei propinqua. Et così si ritrovorno tucti quelli famosi homini insiemi di unità concordia, per la qual cosa conphresi la causa che faceva stare quella regina con la mente sì suspesa. Onde io anchora insiemi con loro hebbi di ciò grandissima alegrezza et a quello homo factomi vicino, humilmente a llui racomandami et da quello fui con leto animo et benigno aspecto acceptato. Et tanto loro
163r conditione et dominio mi piaque che | omni minima administratione che in quella corte havuta havessi volentieri vi saria dimorato. Ma la abundantissima alegreza et gaudio occupava tanto loro animi che per allora non potetti mio nome manifestarli, ma spero che anchora da quelli nel numero di loro ministri et servitori sarò collocato, d'il che allo inmenso Idio riferirò gratie inmortali a quello per sua gratia fo humile prece che di tale dono, benché inmerito, mi faccia degno.

E avete, Magnifico Juliano, nelle tre sopradicte parte la forza del numero quanto sia potente et quanta virtù in epsa si concluda. Et comme el suo procedere è infinito, così la sua virtù è infinita. Et per essere dicta virtù infinita et il nostro intellecto è finito et *a finito ad infinitum nulla est proportio*, comme dice Aristotele, non può el nostro intellecto sua infinita virtù conphrendere, possemo bene quanto si extende la nostra fragile memoria in qualche parte di
163v epsa virtù conphrendere, ma non in tucto. Et comme dice el vulgare | proverbio: tucti li homini sanno tucte le cose ma omni homo non sa tucte le cose. Credo bene che quando si acozassino insieme multitudine di homini in tale

8 sui] F sua 14 lavatosi] F levatasi 15 se dimorassi] F dimorassi 30 essere dicta] F essere la sua 36 in] F di

scientia docti et eruditi che se non in tucto, in grandissima parte se intenderebbe sue mirabile forze. Ma uno piccolo vermicello, comme sono io, non che tucte, ma una delle sue minime parte non mi reputo sapere il principio, perché, sì comme è dicto, non vi si può dare proportione, sono ingegnatomi di quelle poche forze che a me sono note, darne quella vera notitia che a me è suto possibile et con quella brevità che ho potuto. Et sono tucti casi piacevoli al parere mio et da dilectare molto la mente perché sono absoluti. Ma quando uno di questi casi sarà proposto a uno che non habbi così notitia del modo per il quale si absolvano, non li sarà né di piacere né di dilecto, perché lo farà molto bene fantastichare et pigliare et posare la penna più di dua volte. Et io ne so dare buona testimonianza che vi è tale caso et dimostra|tione in su questo libretto che più di 6 volte per ritrovarla me ha 164r facto suadare le tempie. Et mostrasi anchora la forza dell'arismetrica, cioè del numero, ché l'arismetrica va per numero et la geometria va per misura, casi più fantastici et più rumatici che sono questi et benché sieno soluti niente di mancho, sono di grandissimo fastidio. Et maxime a quelli che in questa scientia non sono bene bene pratici et experti. La geometria anchora dimostra in sé mirabilissima forza et virtù, comme Euchlide Megharese, philosopho dignissimo, nelli sua quindici libri apertamente dimostra, le quali forze vanno per via di linee rette et curve, circuli, quadrati et tranguli et altre figure corporale. Et sono di grandissimo piacere et consolatione quando sono soluti a ciascheduno. Ma questa forza et virtù per adesso non intendo in questo nostro tractato essere. Solo in questa quarta parte voglio mostrare, comme nella prima parte dissi, li modi di misurare una torre con l'occhio solamente et da misurare una | distantia di uno fiume o simile con l'occhio et qualche 164v altra gentileza, le quali cose non passeranno senza vostra grandissima consolatione, comme poterete vedere. Le quali forze tucte derivano dalla virtù della geometria et per via lineale si fa molte mirabili operationi. Et stendesi la forza et virtù geometrica a condurre le aque et alla pictura et alla prospectiva et sculture et molte mirabili arti da questa scientia procedano. Hora attenderemo andare innanzi alla nostra opera. Et prima diremo, col nome di Dio, così.

Prima investigatio da misurare una torre

Diremo dunque in questa nostra quarta et ultima parte qualche cosa dilectevole et di piacere et solum reduco a memoria a Vostra Magnificentia che queste

4 sì comme] F comme 6 che ho] F ho 15 rumatici] K, F rinnatici

cose sono materia sottile et difficile et non si possono apertamente dire che non paino difficile et a volerle intendere bisogna mettervi tucto lo ingegno et fantasia et queste cose sono molto rare et da pochi o da quasi niuno sino a 165r hoggi date in pubblico. Tamen mi sforzerò totis viribus | dirle più aperte che a me fia possibile. Daremo dunque in questo primo capitulo el modo di misurare una torre con l'occhio solamente. Bisogna essere in luogho che risponda in piano con la torre insiemi et se non è, si può porre in piano facilmente. Se voi fussi, verbi gratia, in sul canto di vacchereccia et vogliate sapere el Palagio delli Signori quanto sia alto senza altrimenti misurarlo solo con l'occhio fate così: fermatevi in su dicto canto et ficchate in terra quivi uno dardo dinanzi a voi che stia inmobile et fermo overo potete torre una asta ferrata a simile artificio deputata, la quale bene si possa in terra fermare. Et poi schostatevi da questa hasta o dardo quanto pare, a duoi o tre o quattro braccia, non importa, et poi mirate la sommità del palagio overo torre che voi volete misurare per el diricto della vostra hasta et, dove vedete che batta el vostro vedere per il dardo, fatevi porre uno pocho di cera et chiameremo questa cera A. Et guardate voi di non vi 165v movere per niente ché movendovi puncto puncto non pote|resti fare el vostro intento et perhò voi sarete inmobile et anchora el dardo sarà inmobile. Facto havete questo senza partirvi del loco, mirate pure per il vostro dardo el piè uno di decto palagio o torre, cioè el luogho dove si farebbe a cominciare a misurare dicto palagio quando lo volessi misurare. Et similiter, dove batte el vostro occhio nel dardo, fate porre una seconda cera et chiamate questa seconda cera B. Poi, col medesimo occhio del medesimo locho mirate per il vostro dardo uno luogho in dicta torre o palagio facile da potersi da voi misurare con la vostra hasta quando vi acosterete alla torre come, verbi gratia, una bucha che fussi in dicta torre overo lo archo della porta o simili cose note. Et così dove batta el vostro occhio nel dardo farete porre una tertia cera et chiaramente questa tertia cera C et per più vostra facilità porremo qui a piè la pictura in forma propria.

Dico che in questo caso, quante volte entra la parte del vostro dardo quale 166r sta fra la cera B e la cera C in quella altra parte che sta infra | la cera A e la cera C, tante volte quella parte del palagio ad voi nota, cioè quella mira bassa atta a potersi da voi misurare et

1 si possono apertamente] F si possono sì apertamente 2 difficile] F obscure 8 vogliate] F voglia 9 fate] F fa 12 poi] K po' 18 fco (facto) scritto sopra un testo depennato. Nel codice F facto aggiunto a margine.

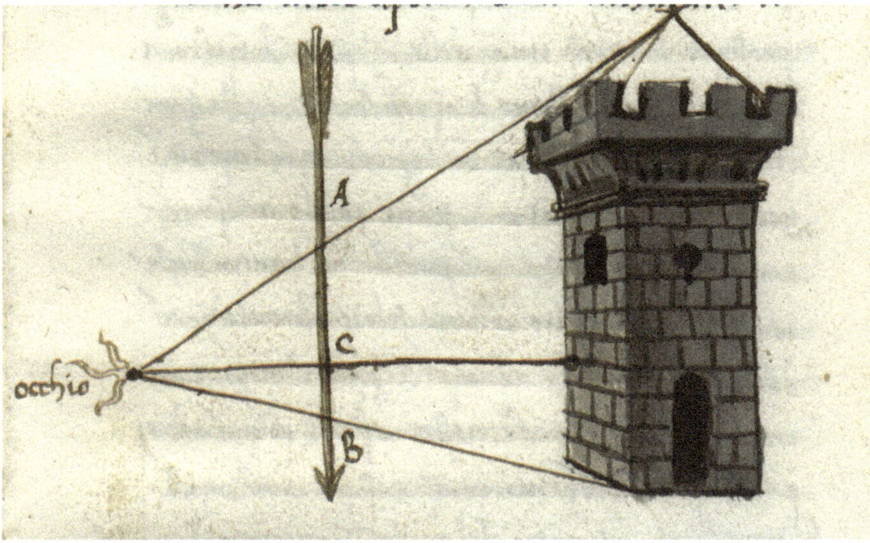

FIGURE 16 Drawing on folio 166r

ad voi nota, entrerrà nella parte superiore della torre ad voi incognita. Et per più vostra chiareza daremo lo exemplo del numero. Diciamo che il palagio delle Signori sia alto braccia 130 et il segno che voi mirate comme lo archo della porta o simili, sia alto braccia 10. Troverrete nel vostro dardo simile ragione, cioè che comme quella parte ad voi nota della torre, entra | nel resto del palagio 12 volte 166v et in sé è una delle 13 parte del dicto palagio, così la parte del dardo CB entrerrà nella parte AC 12 volte et questo BC sarà una delle 13 parte di tucto AB. Et mai vi fallirà la predicta regola pureché sempre vi troviate con li piedi et con l'occhio in uno medesimo stato, siavi facile a misurare omni grande alteza et quando non havessi hasta o dardo, potrete fare con uno filo piombinato facendolo pendere innanzi ad voi et segnare in epso le mire vostre con 3 paternostri o perle, comme ad voi parrà più a proposito.

Secunda investigatio del modo di misurare una torre con l'occhio solamente per modo più difficile

Se fussi una torre della quale voi non potessi havere nota alcuna parte ma bene potete andare sino alli pie' di dicta torre et volessi con l'occhio solamente misurarla, anchora a questo vi darò uno legiadro modo et ordine. Scostatevi dalla prenominata torre o palagio quanto pare ad voi con modo potere bene vedere la sua sommità et di|nanzi ad voi ficcate uno dardo overo hasta che tucto è 1 167r o quando non havessi niuna delle dicte cose, pigliate uno filo piombinato et fatelo pendere davanti a voi et con perle o paternostri signate li puncti et poi vi scostate da dicto dardo quanto vi pare et ponete l'occhio vostro giù in terra basso et di quivi mirate la cima della torre et dove vedete che nel dardo batta el razzo della cima della torre, quivi farete porre una cera et chiamerete la cima del dardo A et el piè del dardo B et questa cera che vi ponete si chiamerà C et l'occhio vostro si chiamerà D come giù a piè vedete la pictura in forma propria.

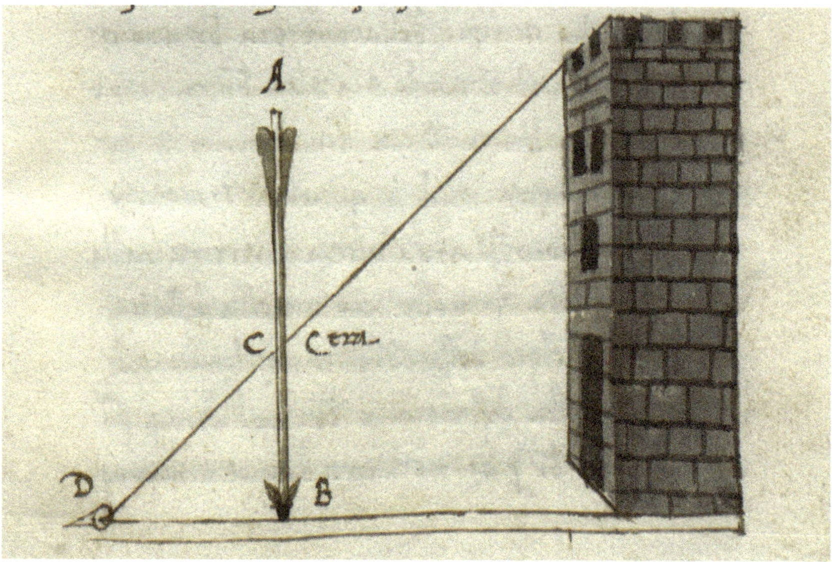

FIGURE 17 Drawing on folio 167r

167v In questo, dico, troverrete che la parte del dardo quale è fra C e B enterrà nella distantia che è fra DB, cioè nella distantia che è dall'occhio vostro et el piè del dardo, quante volte l'alteza di tucta la torre entra nella distantia che è fra l'occhio vostro et la torre. Adunque voi misurate quante volte CB entra in DB et tante volte direte che la torre entri in tucta la distantia che sia fra l'occhio vostro et la torre et misurerete dicta distantia et partiretela per le volte che entra CB in DB et verranne l'altezza di dicta torre. Eccho lo exemplo materiale: sia la distantia che è fra voi, cioè fra l'occhio vostro et il piè della torre sia braccia 400 et la mira del vostro dardo sia alta braccia 2, cioè CB sia 2 braccia et DB sia distante 8 braccia, dunque CB entra in DB 4 volte. Troverrete che la torre, cioè la sua alteza entrerrà nella distantia decta 4 volte, cioè nella distantia fra l'occhio, el piè della torre et noi; havemo dicto che la dicta distantia è braccia
168r 400. Dunque partiamo 400 braccia per 4, ne verrà 100 et troverai | che la torre sarà alta braccia 100 apuncto. Et mai potrete errare servando questo ordine.

6 enterrà F] entrerrà

Sono alcuni che tengono questo ordine, cioè che si apressano tanto alla torre che stando a giacere roverscio et tocchando con li piè el dardo ficto in terra comme di sopra la mira vostra adirittura alla cima della torre batte el raggio nel dardo alto apuncto quanto sarà l'occhio vostro se voi fussi ritto. Et questa regola io per me la confirmo: perché è per la prima data regola, comme è dicto, perché entrando CB in DB una volta lo spacio fia l'occhio vostro et el piè della torre sarà apuncto la alteza di dicta torre. Dunque tanto sarà dal piè della torre all'occhio vostro quanto sarà dal piè della torre alla sua cima. Et è questa medesima regola che di sopra ho data, ma questa è maggior faticha che avanti che lui habbi trovato el logho che a tale misura li responda col primo modo, voglio havere misurato 2 volte dicta torre pure omni uno pigli a sua comodità quello vuole et quello più li dilecta.

Tertia investigatio di uno altro modo di misurare similiter una altezza di una torre con l'occhio 168v

Leggesi nelli antiqui volumi uno modo di misurare similiter una alteza di una torre, el quale è modo verissimo et utile et fassi in questo modo et forma, cioè: toglie uno specchio o quando non havessi specchio vi servirà uno infreschatoio pieno di aqua chiara che omni uno farà el medesimo effecto. Et ponetelo in locho piano donde facilmente si veggha la cima di decta torre et poi vi scostate da dicto specchio a pocho a pocho, tanto che voi veggiate in dicto specchio overo in decta aqua la cima di decta torre apuncto, tenendo sempre el viso volto verso la torre e lo specchio et comme vedete la cima della torre nella superficie dell'aqua et voi vi fermate et troverrete che quante volte lo spatio che è fra ll'occhio vostro et li piè vostri entra nello spatio che è fra li piè vostri et lo specchio tante volte la alteza della torre entra nello spatio che è fra llo spechio et el piè di dicta | torre comme qui da piè vedi in figura. 169r

Sia la cima della torre A, el piè suo B et lo specchio sia C et il vostro occhio sia D et li vostri piedi sieno E. Troverrete che tale è DE ad EC quale è AB ad CB. Ecco lo exemplo del numero: sia alta la torre braccia 100 et lo spatio fra lo specchio et el piè della torre sia braccia 500. Troverrete che lo spacio che è fra lli vostri piedi et el vostro occhio cinque volte entra nello spatio che è fra li vostri piedi et lo specchio et mai questa regola fallirà et è brieve et verissima a iuditio di tucti.

Quarta investigatio a misurare una distantia di uno fiume o simile larghezza 169v

Quando voi non vi potessi acostare alla torre, cioè che vi fussi in mezo uno fiume o simili purché voi veggiate el piè di dicta torre vi bisogna trovare modo di sapere quanto sia dalli vostri piedi al piè di dicta torre perché se voi saprete misurare questo spatio bene con la sopradicta misura, saprete poi ritrovare la sua alteza facilmente. Per sapere la sopradicta distantia vi servirà questo modo che qui di sotto darò a misurare una distantia di uno fiume overo di uno prato

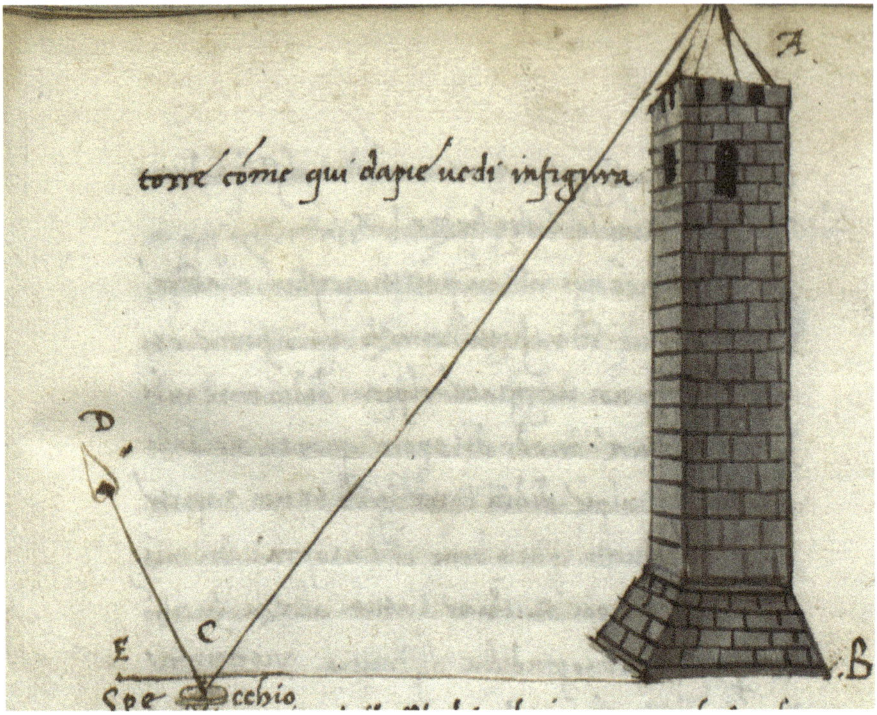

FIGURE 18 Drawing on folio 169r

et omni altra distantia et maxime quando non sia troppo grande. A misurare una distantia grande bisogna altro modo. Volendo dunque misurare la distantia di uno fiume essendo in su una delle ripe, farete in questo modo: ponetevi con li piedi vostri in luogho piano et fermo et quivi ficchate in terra uno dardo fermo, inmobile et questo dardo si chiama el suo capo A, el suo piè B. In questo
170r dardo proprio all'alteza ove batte l'occhio vostro | et voi ponete una cera et chiamasi questa cera C et poi vi scostate da questo dardo quanto aprite le braccia o più o meno quanto vi pare pure in luogho piano et conrispondete a lluogho ove ficchaste el primo dardo et voi ne ficcate uno altro el cui capo è D et il ferro E. Et in questo dardo similmente all'alteza apuncto del vostro occhio ponete una cera et chiamasi questa cera F. Ponete l'occhio vostro apunto a questa cera F et mirate per dirictura del dardo AB qualche cosa nota di là dal fiume. O, se avete a misurare una torre, mirate el piè di dicta torre, ma, quando non vi sia la torre, mirate uno albero grande o uno sasso o uno cespuglio o simili cose note. Et chiamasi questa cosa nota G et dove mirando per il dardo AB el vostro raggio taglia el dicto dardo, quivi ponete una altra cera et chiamasi questa cera H

7 aprite le braccia] F aprite

comme qui a piè vedete in margine signata la pictura. In questo caso dico che
se voi misurerete lo spatio di una cera all'altra del primo dardo, cioè del dardo
AB, lo spatio | CH quante volte entra nello spatio che è fra uno dardo et l'altro, 170v
cioè nello spatio CF troverrete che HB entra in BG, cioè nello spatio che è fra voi
et lo albero che voi mirasti di là dal fiume. Misurerete dunque quante volte HC
entra in FC et tante volte arguirete che HB entrerà in BG et mai poterete errare
seguendo questo ordine.

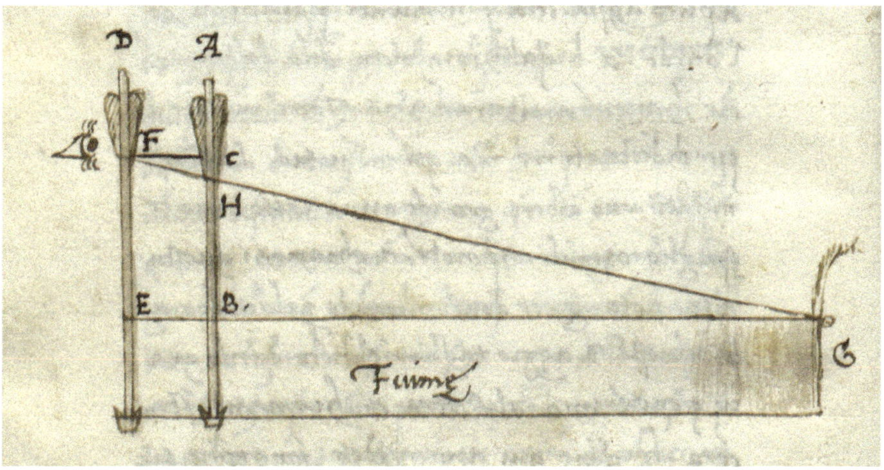

FIGURE 19 Drawing on folio 170v

Ecchovi lo exemplo: sia largho el fiume braccia 40 et lo spatio HB sia braccia
2. Troverete che CH entrerà in FC venti volte et el fiume sarà | largho 20 volte 171r
2 braccia, cioè 2 via 20 fa 40 et tanto sarà largho el fiume et così similiter se
CH entrerà in FC 30 volte troverrete anchora che FE entrerà pure in EG 30 volte
et mai questa regola potrà fallire et con questo modo misurerete una torre che
fussi di là dal fiume et mille belle fantasie che sono mirabilissime nel conspecto
quasi universale delli homini perché a volere fare queste misure bisogna havere
el capo a botegha et tenere il cervello fermo.

Investigatio quinta: uno altro modo a misurare una distantia

Fassi uno altro modo ad investigare una largheza di uno fiume quando el paese
sia piano che è più facile et più aperta et fassi così: ficchate uno dardo AB in
terra et alla dirittura del vostro occhio in questo dardo AB ponete una cera si-
milmente et chiamasi questa cera C. Et poi vi scostate da questo dardo 2 braccia
incircha et ficchate uno altro dardo adirictura a filo di quello et chiamasi questo
dardo DE et similmente | apuncto alla alteza del vostro occhio in questo dardo 171v
DE ponete una altra cera et chiamate questa cera F. Ponete el vostro occhio

comme nella superiore alla cera F et mirate per il dardo AB el loco G, cioè el loco che sta di là dal fiume o albero o sasso o quello si sia et dove el vostro vedere taglia dicto dardo AB, quivi ponete una altra cera et chiamasi questa cera H. Poi misurate nel dardo AB quanto è dalla cera H alla cera C et ponete una cera nel dardo DE che sia sotto la cera F per quanto spatio quanto sta presso la cera H alla cera C apuncto né più né mancho, et chiamasi questa cera I. Et poi ponete el vostro occhio al lato alla cera C voltando le spalle al fiume et mirate la cera I et vedete dove el vostro vedere batte là discosto in terra. Et dove batte fate porre uno saxo o uno segno o qualche cosa da tenere a mente. Et chiamasi questa cosa K comme qui a piè vedete picto in figura.

Dico che in questo caso troverrete che tanto sarà dal signo K per insino al
172r dardo AB quanto sarà | dal dardo DE al signo G, cioè a quello signo che sta di
là dal fiume. Et questa è misura certissima et infallibile et con questa si potrà misurare una distantia grande quando el paese sia piano.

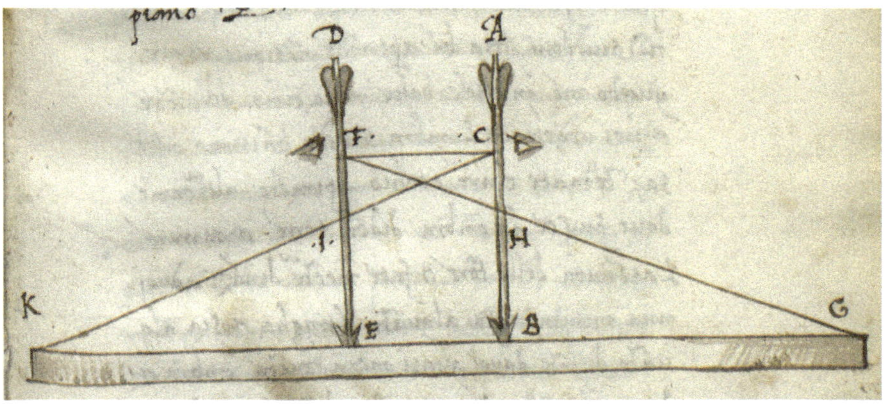

FIGURE 20 Drawing on folio 172r

Sexta investigatio: altro modo a misurare una torre

Fanno li antiqui mathematici certi modi da misurare una torre el quale anchora non è da sprezare et va per via di proportione. Ma questo modo non serve se non quando è la spera del sole, ma quando fusse nuvolo non vi servirà et perhò
172v non è troppo stimata. Tamen la voglio inducere | in questa nostra opera che
qualche volta vi potrà servire et è misura chiara. Et fassi con l'ombra del sole, cioè quando l'ombra del sole darà nella torre in modo che in terra apparischa sua ombra manifesta. Et quando la terra sarà in luogho che l'ombra non si vedessi stesa bene, apertamente non si può usare questo modo. Quando el sole

2 dove el vostro vedere taglia] F dove è el locho vedere tale

batte nella cima della torre et voi vedete che l'ombra si vede benissimo distesa, fermate el vostro occhio apuncto al termine dove finisce la ombra di dicta torre et misurate la sommità della torre. Et fate mectere dinanzi ad voi una misura nota alquanto lungha et sia alquanto distante da voi quasi meza dicta ombra. Et dove el vostro vedere taglia dicta lancia, segnate o fate signare o veramente la farete porre in modo che la sommità di dicta lancia apuncto tocchi el raggio del vostro occhio comme qui in figura propria vedete signato.

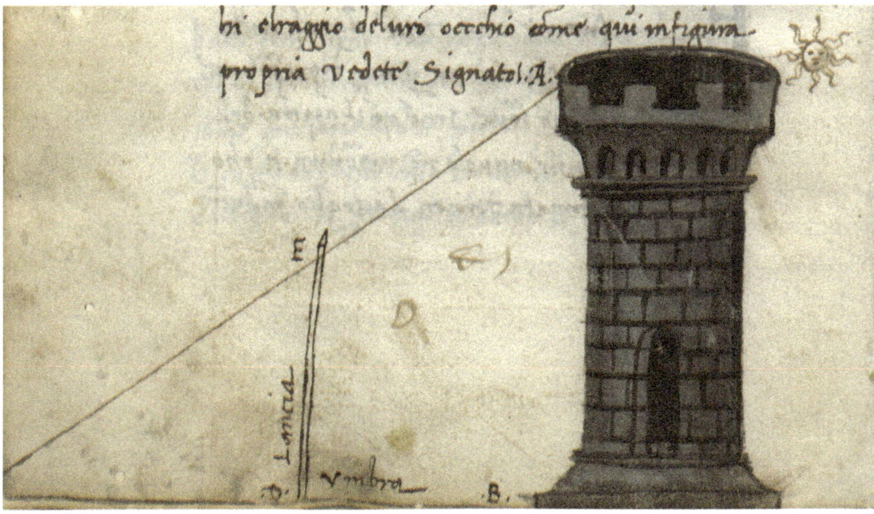

FIGURE 21 Drawing on folio 172^{v}

Sia adunque la torre AB et la ombra di epsa torre CB. L'occhio vostro sarà in sul 173^{r}
puncto C et la lancia sirà el piè che è el primo in sulla ombra sirà D et il capo suo che taglia el raggio CA sirà E comme vedete dipincto. Dicono dunque li antiqui scriptori che tale sarà DC ad DE quale sarà BC ad BA. Dunque se voi misurerete quante volte CD entra in DE troverrete che alla medesima ragione enterrà CB in BA. Ma questo modo io per me non l'ho mai usato tamen qui non lo ho voluto preterire et perhò non me extenderrò troppo in epso dare exempli. Basta vi ho dato suo ordine et modo di poterla usare quando volessi farne experimento.

Septima investigatio a misurare una profundità d'uno pozzo o simili

Con questi modi et regole di sopra recitati potrete misurare omni et qualunche alteza a occhi solamente et comme si misura le altezze potrete con questa data regola misurare le profundità. Ma a ciò meglio Vostra Illustrissima Magnificen-

8 sia] K si 10 sirà E] F sirà 11 DE] F D

173^{v} tia | le possa intendere ne daremo uno exemplo. Se voi volessi misurare quanto
sia profundo uno pozzo o citerna o simili profundità, terrete questo ordine:
traversate una hasta dentro nel pozo quanto più giù potete con mano a giugnere
et fermatela che la stia da sé ferma in modo che sia diametro di epso pozzo.
Ponete poi l'occhio vostro in sul'orlo del pozzo in modo che 'l vostro occhio
venga a piombo sopra el capo di dicta hasta et bisogna che quello tale luogho
sia per modo che voi possiate vedere el cavato di decto pozzo sino all'aqua et col
vostro occhio mirate lagiù all'aqua lo orlo della sua superficie in modo inverso
l'altra parte della hasta, cioè chiamasi el capo della hasta che è sotto l'occhio
vostro apuncto A et el capo di dicta hasta a voi riscontro si chiami B et l'occhio
vostro si chiami C et la superficie dell'aqua si chiami D. Voi dunque mirate con
l'occhio inverso el capo B della hasta et mirate la superficie dell'aqua nel locho
D et dove el vostro occhio taglia l'hasta quivi apuncto fate porre uno signo, cioè
174^{r} una cera et | chiameremo questa cera E. Et notate che quando non haveste hasta
similmente potreste torre uno filo et quello con due aguti conficchare al muro
di qua et di là. Hora, per più vostro chiaro exemplo vi porrò qui da piè la pictura
in forma propria.

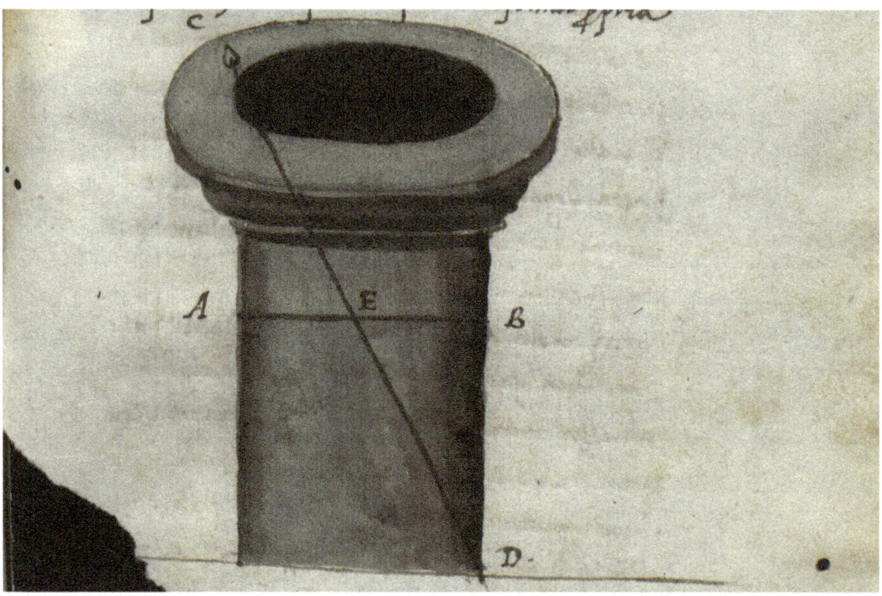

FIGURE 22 Drawing on folio 174^{r}

Troverrete dunque che quante volte AE entra in AC, cioè quante volte lo spa-
174^{v} tio che sta fra l'hasta AE entra nello spatio che sta fra el capo de|l'hasta A et el
vostro occhio C tante volte tucta l'hasta enterrà nello spatio tucto del pozzo.

Misurerete dunque quante volte AE entra in AC et poi arguirete che dallo orlo del pozzo all'aqua sia tante haste comme se la larghezza del pozzo, cioè AB, sia braccia 3 et el pozzo sino all'aqua sia braccia 18 cupo dunque la hasta AB misura 6 volte tucto el pozo. Così troverrete nella vostra hasta la medesima ragione, cioè che AC enterrà in CA 6 volte et mai vi fallirà questa regola.

Potrei molti altri exempli a Vostra Illustrissima Magnificentia allegare, ma questi soprascripti bastino per al presente al vostro mirabile ingenio. Et con queste forze et virtù numerali sino ad hora recitate si può venire a molti varii et innumerabili effecti che qua non sono inclusi. Ma io mi sono sforzato in questa mia operetta dire cose di piacere et dilecto et non da dare fastidio perché le cose sottile a chi non le ha molto bene alla mente et chiare non passano senza grandissima ansietà et fastidio | di mente. Et, comme dice Archimede, homo in 175r questa scientia molto in verità peritissimo: "*mathematicha et logica requirunt totum hominem*". Non sono queste due scientie et anchora buona parte delle altre da dire: "io vi attenderò quando mi avanzerà tempo" ché uno che in simile modo vi attende mai verrà perfecto in niuna scientia. Et perhò che io so che Vostra Magnificentia Illustrissima è circa maiora occupato, li casi quali ho in questa mia opera decti, mi sono sforzato dirli asboluti quanto mi è suto possibile, aciò Quella non habbi in simile cose a perdere tempo, ma solo quando, per refrigerare alquanto lo animo, levato via le mense insieme con le vostre Carissime et Magnifiche sorelle Madonna Lucretia, quale sempre è stata nostro unico refugio, et Madonna Madalena, la quale per sua innata humanità et immensa gentilezza si degnò fare sì gratissima acoglienza et optima conpagnia a mia observanda Madre che fu ultra modum et Madonna Contessina di questi | nostri ludi, casi et dimostrationi di sopra dicte, con quelle pigli alquanto 175v di piacere et consolatione. Alla quale Vostra prefata Illustrissima Magnificentia del continuo mi racomando et priegola che del suo fido et sincero servitore Piero qualche volta si degni ricordarsi. Et allo inmenso et sommo et immortale et altitonante Idio referite le debite et conveniente gratie et Quello divotamente priegho che in felicissimo stato Vostra prelibata Magnificentia mantenere et conservare si degni. Daremo adunque alla nostra operetta debito et felice fine. *Itenus laus Deo omnipotenti qui nos ob eius infinitam potentiam atque misericordiam ad celestia regna perducere dignetur. Finis.*

29 debite] F dite 30 priegho] F pregato

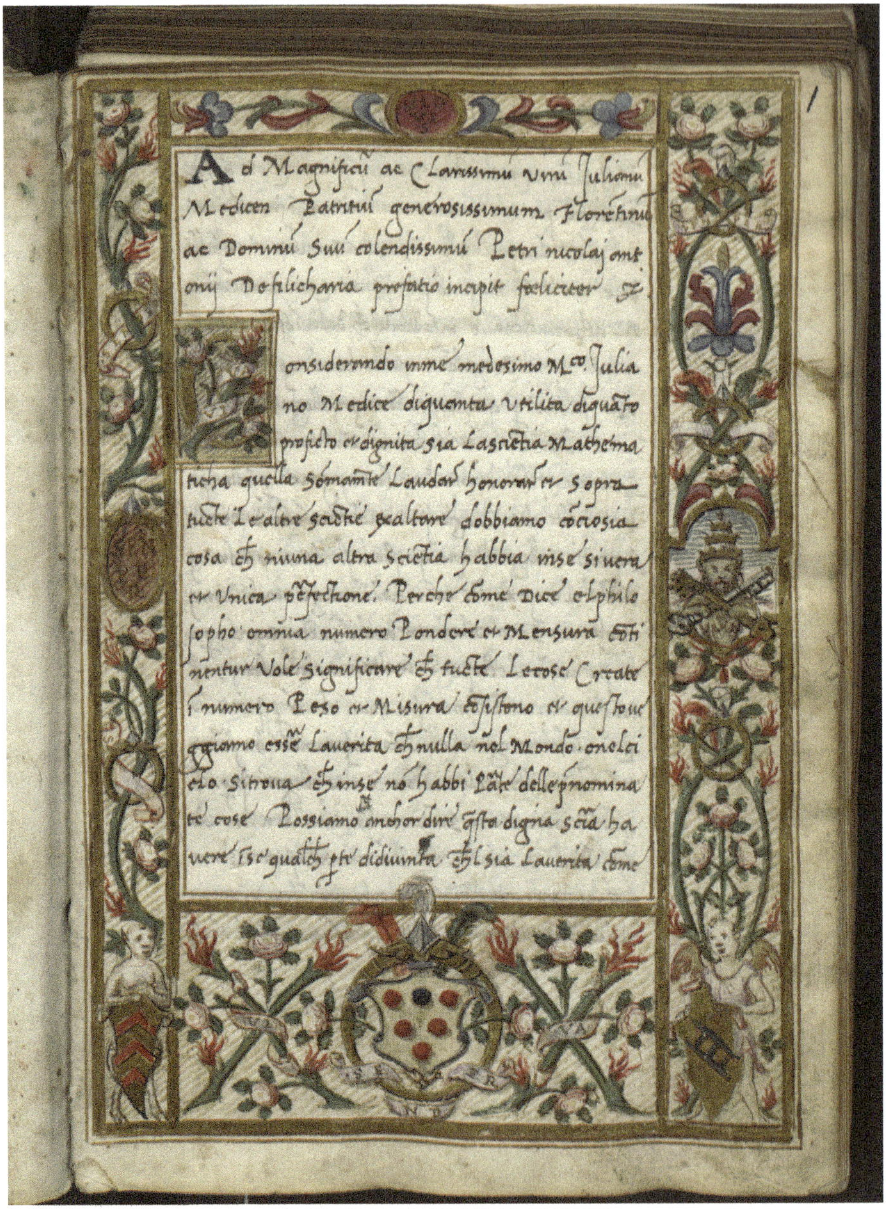

FIGURE 23 First folio of the manuscript Ital. Quart. 48

Bibliography

1 Manuscripts of the *Giuochi mathematici*

Filicaia da, Niccolò, *Giuochi mathematici*, Biblioteka Jagellońska, Ital. Quart. 48 (first quarter of the 15th century).

Filicaia da, Niccolò, *Giuochi mathematici*, Biblioteca Magliabecchiana, XI 15 (first quarter of the 15th century).

2 Studies

Alcuino da York 'Manuale di storia della filosofia medievale on-line', Università di Siena Facoltà di lettere e filosofia. http://www3.unisi.it/ricerca/prog/fil-med-online/autori/htm/alcuino_york.htm. (Consulted on 5 February 2020).

Agostini, Amedeo, '*De viribus quantitatis* di Luca Pacioli', *Periodico di Matematiche*, Vol. IV (1925), 165–192.

Alberti, Leon Battista, 'Delle piacevolezze delle Matematiche', in *Opuscoli morali di Leon Batista Alberti gentil'huomo firentino: ne' quali si contengono molti ammaestramenti, necessarij al viver de l'huomo, così posto in dignità, come privato. Tradotti, et parte corretti da m. Cosimo Bartoli*, (Venezia: Francesco Franceschi, 1586), 225–255.

Alberti, Leon Battista, '*Ludi rerum mathematicarum*', in *Opere volgari*, ed. by Cecil Grayson, (Bari: Laterza, 1973), III, 133–173.

Alberti, Leon Battista, *Ludi matematici*, ed. by Raffaele Rinaldi (Milano: Guanda, 1980).

Arrighi, Gino, 'Il "Libro dicto giuochi mathematici" di Piero di Nicolao d'Antonio da Filicaia', *Atti Fondazione Ronchi*, 26 (1) (1971), 51–61.

Bembo, Pietro, *Lettere*. Ed. by Ernesto Travi, Collezione di opere inedite o rare. (Bologna: Commissione per i testi di lingua, 1987).

Bianco, Monica, 'Predicazione e letteratura nelle trascrizioni di Antonio da Filicaia', in *Letteratura in forma di sermone: i rapporti tra predicazione e letteratura nei secoli XIII–XVI: atti del seminario di studi, Bologna 15–17 novembre 2001*, ed. by Ginetta Auzzas, Giovanni Baffetti, Carlo Delcorno, (Firenze: L.S. Olschki, 2003), 233–245.

Blasco Ferrer, Eduardo, *Primo corso di linguistica italiana* (Firenze: Franco Cesati, 2015).

Boncompagni, Baldassare, 'Intorno alle vite inedite di tre matematici scritte da Bernardino Baldi (Giovanni Dank di Sassonia, Giovanni de Lineriis e Fra Luca Pacioli da Borgo San Sepolcro)', in *Bullettino di Bibiliografia e Storia delle Scienze Matematiche e Fisiche*, XII (1879), 352–438.

Bressanini, Dario, Toniato, Silvia, *I giochi matematici di fra' Luca Pacioli: trucchi, enigmi e passatempi di fine Quattrocento* (Bari: Dedalo, 2011).

Buonaccorsi, Biagio, *Diario dall'anno 1498 all'anno 1512 e altri scritti*, ed. by Enrico Niccolini (Roma: Istituto Storico Italiano per il Medio Evo, 1999).

Castiglione, Baldassarre, *Lettere famigliari e diplomatiche*, ed. by Guido La Rocca, Angelo Stella, Umberto Morando, 3 Vols (Torino: Einaudi, 2016).

Ciocci, Argante, *Luca Pacioli: la vita e le opere* (Sansepolcro: University Book, 2017).

Conterio, Annalisa, 'Il testo e la sua storia', in *Trattato di partita doppia*, Luca Pacioli, ed. by Annalisa Conterio (Venezia: Albrizzi, 1994), 45–50.

D'Amore, Bruno, 'Leon Battista Alberti ed i suoi *Ludi rerum mathematicarum*', *Il Carobbio*, XXX (2005), 61–66.

Delcroix, Rita, *Giuliano de' Medici: il crepuscolo del Rinascimento* (Firenze: Camunia, 1996).

Devlin, Keith, *I numeri magici di Fibonacci* (Milano: Rizzoli, 2013).

Egmond, Warren Van, *Practical mathematics in the Italian Renaissance: a catalog of Italian abbacus manuscripts and printed books to 1600* (Firenze: Istituto e Museo di storia della scienza, 1980).

Franci, Raffaella, 'Giochi matematici in trattati d'abaco del Medioevo e del Rinascimento', in *Atti del Convegno nazionale sui giochi creativi: Siena 11–14 giugno 1981*, (Siena: Tipografia senese, 1981), 18–43.

Franci, Raffaella, 'Jealous Husbands Crossing the River: A Problem from Alcuin to Tartaglia' in *From China to Paris: 2000 years transmission of mathematical ideas*, ed. by Yvonne Dold-Samplonius, Doseph W. Dauben, Menso Folkerts, Benno van Dalen, (Stuttgart: Steiner, 2002), 289–306.

Franci Raffaella, ed. by, *Alcuino di York, Giochi matematici alla corte di Carlomagno* (Firenze: Edizioni ETS, 2005).

Frosini, Giovanna, 'Firenze' in *Città italiane, storie di lingue e culture*, ed. by Pietro Trifone, (Roma: Carocci, 2015), 203–224; 216–221.

Frosini, Giovanna, 'Lingua', in *Enciclopedia Machiavelliana*, 3 Vols (Roma: Istituto della Enciclopedia Italiana Treccani, 2014), II, 720–732.

Franco, Matteo, *Lettere*, ed. by Giovanna Frosini (Firenze: Accademia della Crusca, 1990).

Furlan, Francesco, 'In margine all'edizione degli *Ex ludis rerum mathematicarum*: ossia osservazioni e note per l'edizione di un testo scientifico e delle sue figure', *Revue d'histoire des sciences* 59 (2) (2006), 197–217.

Gar, Tommaso, 'Documenti risguardanti Giuliano de' Medici e il pontefice Leone X', *Archivio Storico Italiano*, Append. 1 (1842–1844), 299–324.

Garfagnini, Gian Carlo, 'Tra politica, clientele e senso dello stato: Bartolomeo Scala', *Annali del Dipartimento di Filosofia* (*Nuova Serie*), XV (2009), 109–130.

Geymonat, Ludovico, 'Prefazione', in *Ludi matematici*, ed. by Raffaele Rinaldi (Milano: Guanda, 1980).

Guerraggio, Angelo, *Matematica* (Milano: Egea, 2012).

Guicciardini, Francesco, *Storie fiorentine dal 1378 dal 1509*, ed. by Roberto Palmarocchi (Bari: Laterza, 1931).

Jayawardene, Sugathadasa A., 'Luca Pacioli' in *Dictionary of Scientific Biography*, ed. by Charles Coulston Gillispie, (New York: Scribners, 1974), Vol. 10, 269–272.

Kostecki, Paweł, 'Krótka historia matematyki', https://www.fuw.edu.pl/~kostecki/histmat.pdf. (consulted on 2 January 2020).

Kristeller, Paul Oskar, *Iter Italicum: a Finding List of Uncatalogued or Incompletely Catalogued Humanistic Manuscripts of the Renaissance in Italian and other Libraries* (Leiden: Brill, 1963) http://www.IterGateway.org (consulted on 13 September 2019).

Kristeller, Paul Oskar, *Iter Italicum: a finding list of uncatalogued or incompletely catalogued humanistic manuscripts of the Renaissance in Italian and other libraries*. (London; Leiden: Brill, 1963).

McManamon, John M., 'Marketing a Medici Regime: The Funeral Oration of Marcello Virgilio Adriani for Giuliano de' Medici (1516)', *Renaissance Quarterly*, 44 (1) (1991), 141. https://doi.org/10.2307/2862404.

Manni, Paola, 'Ricerche sui tratti fonetici e morfologici del fiorentino quattrocentesco', *Studi di Grammatica Italiana*, VIII (1979), 115–171.

Manni, Paola, 'Dal toscano all'italiano letterario', in *Storia della lingua italiana*, ed. by Luca Serianni, Pietro Trifone, (Torino: Einaudi, 1993) III, 321–342.

Mattesini, Enzo, 'La lingua di un trattato di scacchi attribuito a Luca Pacioli', *Contributi di Filologia dell'Italiana Mediana*, XXI (2007), 47–78.

Mattesini Enzo, 'Scavi lessicali nel *De viribus quantitatis* di Luca Pacioli', in *Pacioli 500 anni dopo, Atti del Convegno di Studi, Sansepolcro, 22/23 maggio 2009*, ed. by Enrico Giusti, Matteo Martelli (Sansepolcro: Centro Studi Mario Pancrazi, 2010) 139–169.

Medici, Giuliano de', *Poesie*, ed. by Giuseppe Fatini (Firenze: Vallecchi, 1939).

Mercanti, Fabio, Paola Landra, '"Ludi matematici" di Leon Battista Alberti'. *Eiris*, 2 (2007), 15–47.

Montebelli, Vico, 'I giochi matematici del *De Viribus Quantitatis*', in *Luca Pacioli e la matematica del Rinascimento*, ed. by Enrico Giusti e Carlo Maccagni (Firenze: Giunti, 1998).

Morison, Stanley. 1933. *Fra Luca de Pacioli of Borgo S. Sepolcro* (New York: The Grolier Club, 1993).

Nardi, Bruno, *Saggi sulla cultura veneta del Quattro e Cinquecento* (Padova: Antenore, 1971).

Pacioli, Luca, *Divina proportione* (Venezia: Paganino Paganini, 1509).

Pacioli, Luca, *Trattato di partita doppia*, ed. by Annalisa Conterio (Venezia: Albrizzi, 1994)

Pacioli, Luca, *De viribus quantitatis. Trascrizione di Maria Garlaschi Peirani dal codice n. 250 della Biblioteca Universitaria di Bologna* (Milano: Castello Sforzesco, 1997).

Palermo, Massimo, 'Sull'evoluzione del fiorentino nel Tre Quattrocento', *Nuovi Annali della Facoltà di Magistero dell'Università di Messina*, IX (1991), 131–156.

Palmarini, Luca, 'Aspetti storico-linguistici della denominazione di alcuni oggetti utilizzati a fini ricreativi nel manoscritto cinquecentesco *Giuochi mathematici* di Piero da Filicaia', in *Języki specjalistyczne w ujęciu diachronicznym i synchronicznym*, ed. by Beata Malczewska, Joanna Woźniakiewicz (Kraków: Wydawnictwo Naukowe Uniwersytetu Pedagogicznego 2021).

Palmarini, Luca, Sosnowski, Roman, 'Geometria in un trattato di giochi matematici dell'inizio del Cinquecento', *Romanica Cracoviensia*, 19 (2019), 43–53.

Palmarini, Luca, Sosnowski, Roman, 'Ma l'uovo era veramente di Colombo?: l'attestazione dell'aneddoto nel manoscritto di Piero da Filicaia dell'inizio del Cinquecento', *Cuadernos de Filología Italiana*, 26 (2019), 167–180.

Parenti, Piero, *Storia fiorentina*, ed. by Andrea Matucci (Firenze: Olschki, 1994).

Pietrini, Davide, *L'umanesimo matematico a Urbino dal tardo Rinascimento alla rivoluzione galileiana*, (Urbino: doctoral Thesis, 2019–2020), https://ora.uniurb.it/bitstream/11576/2683371/1/phd_uniurb_285800.pdf (consulted on 18 March 2024).

Ramusio, Giovan Battista, *Primo volume delle navigationi et viaggi nel qual si contiene la descrittione dell'Africa et del paese del prete Ianni* (Venezia: appresso gli heredi di Lucantonio Giunti, 1550).

Ricci, Laura, 'Il lessico matematico della Summa di Luca Pacioli', *Studi di Lessicografia Italiana*, 12 (1994), 5–71.

Rinaldi, Rinaldo, '"FIGURA PRINCIPIS": Rappresentazioni del potere nella letteratura delle corti padane del Quattrocento', *Neuphilologische Mitteilungen*, Vol. 102, 3 (2001), 269–285.

Ruscelli, Girolamo, *Lettere di principi, le quali o si scrivono da principi o a principi o ragionan di principi ... mandato in luce da Girolamo Ruscelli* (Venetia: Giordano Ziletti, 1562–1581).

Sosnowski, Roman, *Origini della lingua dell'economia in Italia: dal XIII al XVI secolo* (Milano: Franco Angeli, 2006).

Sosnowski, Roman, *Manoscritti italiani della collezione berlinese conservati nella Biblioteca Jagellonica di Cracovia (sec. XIII–XVI)* (Kraków: Wydział Filologiczny Uniwersytetu Jagiellońskiego w Krakowie, 2012).

Sosnowski, Roman, 'Tra Firenze e Borgo San Sepolcro. Una ricognizione sulla lingua del testo inedito di *Giuochi mathematici* di Piero da Filicaia', in *Percorsi linguistici tra Italia e Polonia. Studi di linguistica italiana offerti a Stanisław Widłak*, ed. by Elżbieta Jamrozik, Roman Sosnowski (Firenze: Franco Cesati, 2014), 55–63.

Sosnowski, Roman, 'Zagadki matematyczne z Florencji w rękopisie z początku XVI wieku', in *Dawne literatury romańskie: w świecie tolerancji i nietolerancji, zagadek i tajemnic*, ed. by Maja Pawłowska, Tomasz Wysłobocki, (Kraków: Universitas, 2018), 217–226.

Sticco, Maria, *Arte e sincerità* (Milano: Vita e Pensiero, 1940)

Tabacchi, Stefano, 'Giuliano de' Medici', in *Dizionario Biografico degli Italiani* (2009) ed. by Mario Caravale (Roma: Ist. della Enciclopedia Italiana, 2009), Vol. 73.

Taylor, R. Emmett, *No royal road, Luca Pacioli and his times* (Chapel Hill: University of North Carolina Press, 1942).

Ulivi, Elisabetta, 'Luca Pacioli una biografia scientifica', in *Luca Pacioli e la Matematica del Rinascimento*, ed. by Enrico Giusti, Carlo Maccagni, (Firenze: Giunti, 1994) 21–78.

Ulivi, Elisabetta, 'Scuole d'abaco e insegnamento della matematica', in *Il Rinascimento Italiano e l'Europa. Volume v. Le scienze*, ed. by Antonio Clericuzio, Germana Ernst, (Treviso: Angelo Colla Editore, 2008), Vol. 5, 403–420.

Ulivi, Elisabetta, 'Documenti inediti su Luca Pacioli, Piero della Francesca e Leonardo Da Vinci, con alcuni autografi', *Bollettino di Storia delle Scienze Matematiche*, monographic edition, XXIX, 1 (2009), 15–160.

Ulivi, Elisabetta, 'L'ultimo testamento di Luca Pacioli con altri documenti inediti', *Bollettino di Storia delle Scienze Matematiche*, XXXI, 1 (2011), 35–57.

Ulivi, Elisabetta, 'Su Piero di Niccolò di Antonio da Filicaia, autore del Libro di giuochi mathematici', *Bollettino di storia delle scienze matematiche*, XXXIII, 2 (2013), 235–274. https://doi.org/10.1400/213834.

Vagni, Giacomo. 'Per una nuova edizione delle *Rime* di Giuliano di Lorenzo de' Medici', *Filologia italiana*, 13 (2016), 193–224.

Varchi, Benedetto, *Storia fiorentina* (Firenze: Società editrice delle storie del Nardi e del Varchi, 1843), Vol. 2.

Vianello, Vincenzo, *Luca Paciolo nella storia della ragioneria: con documenti inediti* (Messina: Libreria Internazionale, 1896).

Villa, Alessandra, 'Tipologia e funzionamento del sistema della dedica nell'Italia del Rinascimento', *Line@editoriale*, 2 (2010), http://interfas.univ-tlse2.fr/lineaeditoriale/202 (published online on 9 February 2017, consulted on 7 April 2025).

Vitruvius Pollio, Marcus, *Vitruvius iterum et Frontinus a Iocundo revisi repurgatique quantum ex collatione licuit*, ed. by Giovanni Giocondo, (Firenze: Filippo Giunta, 1513).

Wasserzeichen-Informationssystem (*WZIS*) [online], *Piccard-Online* watermark collection. https://www.wasserzeichen-online.de/wzis/index.php (consulted on 7 April 2025).

Yamey, Basil, 'Luca Pacioli, la *Summa* e il *de scripturis*', in *Trattato di partita doppia*, ed. Annalisa Conterio, (Venezia: Albrizzi, 1994) 1–34.

Index